THE CENOZOIC ERA

65 MILLION YEARS

MILLIONS OF YEARS BEFORE THE PRESENT	System / Period	Series / Epoch	Stage — EUROPE	Age — NORTH AMERICA
	QUATERNARY	HOLOCENE		
0.01		PLEISTOCENE	TYRRHENIAN	WISCONSINAN
			MILAZZIAN	SANGAMONIAN
			- - - - - - - - - -	ILLINOIAN
			SICILIAN	YARMOUTHIAN
			EMILIAN	KANSAN
				AFTONIAN
			CALABRIAN	NEBRASKAN
1.8	TERTIARY	PLIOCENE	PIACENZIAN	BLANCAN
			ZANCLEAN	HEMPHILLIAN
5		MIOCENE	MESSINIAN	
			TORTONIAN	CLARENDONIAN
			SERRAVALLIAN / LANGHIAN	BARSTOVIAN
			BURDIGALIAN	HEMINGFORDIAN
			AQUITANIAN	
22.5		OLIGOCENE	CHATTIAN	ARIKAREEAN
				WHITNEYAN / ORELLAN
			RUPELIAN	CHADRONIAN
38		EOCENE	BARTONIAN	DUCHESNEAN
				UINTAN
			LUTETIAN	BRIDGERIAN
			YPRESIAN	WASATCHIAN
54		PALEOCENE	THANETIAN	CLARKFORKIAN
			MONTIAN	TIFFANIAN
				TORREJONIAN
			DANIAN	DRAGONIAN / PUERCAN
65				

THE MESOZOIC ERA

165 MILLION YEARS

MILLIONS OF YEARS BEFORE THE PRESENT	System / Period	Stage EUROPE	Age NORTH AMERICA
65	CRETACEOUS	MAASTRICHTIAN	GULFIAN
		CAMPANIAN	
		SANTONIAN	
		CONIACIAN	
		TURONIAN	
		CENOMANIAN	
100		ALBIAN	COMANCHEAN
		APTIAN — GARGASIAN / BEDOULIAN	
		BARREMIAN	
		HAUTERIVIAN	
		VALANGINIAN	
		BERRIASIAN	
141	JURASSIC	TITHONIAN	
		KIMMERIDGIAN	
		OXFORDIAN	
		CALLOVIAN	
		BATHONIAN	
		BAJOCIAN	
		AALENIAN	
		LIASSIC	
195	TRIASSIC	RHAETIAN	
		NORIAN	
		CARNIAN	
		LADINIAN	
		ANISIAN	
		SCYTHIAN	
230			

Basin and Range

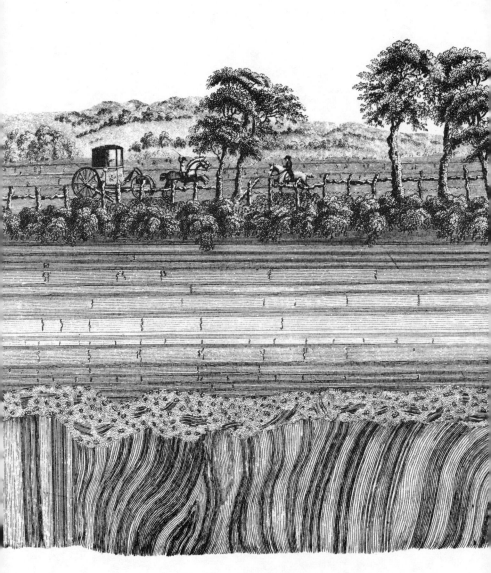

UNCONFORMITY AT JEDBURGH, BORDERS

JOHN
McPHEE

BASIN
AND
RANGE

Farrar · Straus · Giroux

NEW YORK

The text of this book originally appeared in
The New Yorker, and was developed with the editorial
counsel of William Shawn and Robert Bingham

Geological time scale adapted by Tom Funk from the
third edition of F. W. B. van Eysinga's "Geological
Time Table," Elsevier Scientific Publishing Company,
P.O. Box 211, Amsterdam, The Netherlands

Frontispiece by John Clerk, 1787,
courtesy Scottish Academic Press, Ltd., Edinburgh

For Sarah

Basin and Range

The poles of the earth have wandered. The equator has apparently moved. The continents, perched on their plates, are thought to have been carried so very far and to be going in so many directions that it seems an act of almost pure hubris to assert that some landmark of our world is fixed at 73 degrees 57 minutes and 53 seconds west longitude and 40 degrees 51 minutes and 14 seconds north latitude—a temporary description, at any rate, as if for a boat on the sea. Nevertheless, these coordinates will, for what is generally described as the foreseeable future, bring you with absolute precision to the west apron of the George Washington Bridge. Nine A.M. A weekday morning. The traffic is some gross demonstration in particle physics. It bursts from its confining source, aimed at Chicago, Cheyenne, Sacramento,

[3]

through the high dark roadcuts of the Palisades Sill. A young woman, on foot, is being pressed up against the rockwall by the wind booms of the big semis— Con Weimar Bulk Transportation, Fruehauf Long Ranger. Her face is Nordic, her eyes dark brown and Latin—the bequests of grandparents from the extremes of Europe. She wears mountain boots, bluejeans. She carries a single-jack sledgehammer. What the truckers seem to notice, though, is her youth, her long bright Norwegian hair; and they flirt by air horn, driving needles into her ears. Her name is Karen Kleinspehn. She is a geologist, a graduate student nearing her Ph.D., and there is little doubt in her mind that she and the road and the rock before her, and the big bridge and its awesome city—in fact, nearly the whole of the continental United States and Canada and Mexico to boot—are in stately manner moving in the direction of the trucks. She has not come here, however, to ponder global tectonics, although goodness knows she could, the sill being, in theory, a signature of the events that created the Atlantic. In the Triassic, when New Jersey and Mauretania were of a piece, the region is said to have begun literally to pull itself apart, straining to spread out, to break into great crustal blocks. Valleys in effect competed. One of them would open deep enough to admit ocean water, and so for some years would resemble the present Red Sea. The mantle below the crust—exciting and excited by these events—would send up fillings of fluid rock, and with such pressure behind them that they could intrude between horizontal layers of, say, shale and sandstone and lift the country a thousand feet. The intrusion could spread laterally through hundreds of square miles, becoming a broad new layer—a sill— within the country rock.

This particular sill came into the earth about two miles below the surface, Kleinspehn remarks, and she smacks it with the sledge. An air horn blasts. The passing tires, in their numbers, sound like heavy surf. She has to shout to be heard. She pounds again. The rock is competent. The wall of the cut is sheer. She hits it again and again—until a chunk of some poundage falls free. Its fresh surface is asparkle with crystals—free-form, asymmetrical, improvisational plagioclase crystals, bestrewn against a field of dark pyroxene. The rock as a whole is called diabase. It is salt-and-peppery charcoal-tweed savings-bank rock. It came to be that way by cooling slowly, at depth, and forming these beautiful crystals.

"It pays to put your nose on the outcrop," she says, turning the sample in her hand. With a smaller hammer, she tidies it up, like a butcher trimming a roast. With a felt-tip pen, she marks it "1." Moving along the cut, she points out xenoliths—blobs of the country rock that fell into the magma and became encased there like raisins in bread. She points to flow patterns, to swirls in the diabase where solidifying segments were rolled over, to layers of coarse-grained crystals that settled, like sediments, in beds. The Palisades Sill—in its chemistry and its texture—is a standard example of homogeneous magma resulting in multiple expressions of rock. It tilts westward. The sill came into a crustal block whose western extremity—known in New Jersey as the Border Fault—is thirty miles away. As the block's western end went down, it formed the Newark Basin. The high eastern end gradually eroded, shedding sediments into the basin, and the sill was ultimately revealed—a process assisted by the creation and development of the Hudson, which eventually cut out the cliffside panorama of New Jersey as seen across the river from Manhattan: the broad sill, which had cracked, while cooling, into slender columns so upright and uniform that inevitably they would be likened to palisades.

In the many fractures of these big roadcuts, there is some suggestion of columns, but actually the cracks running through the cuts are too various to be explained by columnar jointing, let alone by the impudence of dynamite. The sill may have been stressed pretty severely by the tilting of the fault block, Kleinspehn says, or it may have cracked in response to the release of weight as the load above it was eroded away. Solid-earth tides could break it up, too. The sea is not all that responds to the moon. Twice a day the solid earth bobs up and down, as much as a foot. That kind of force and that kind of distance are more than enough to break hard rock. Wells will flow faster during lunar high tides.

For that matter, geologists have done their share to bust up these roadcuts. "They've really been *through* here!" They have fungoed so much rock off the walls they may have set them back a foot. And everywhere, in profusion along this half mile of diabase, there are small, neatly cored holes, in no way resembling the shot holes and guide holes of the roadblasters, which are larger and vertical, but small horizontal borings that would be snug to a roll of coins. They were made by geologists taking paleomagnetic samples. As the magma crystallized and turned solid, certain iron minerals within it lined themselves up like compasses, pointing toward the magnetic pole. As it happened, the direction in those years was northerly. The earth's magnetic field has reversed itself a number of hundreds of times, switching from north to south, south to north, at intervals that have varied in length. Geologists have figured out just when the reversals occurred, and have thus developed a distinct arrhythmic yardstick through time. There are many other chronological frames, of course, and if from other indicators, such as fossils, one knows the age of a rock unit within several million years, a look at the mineral compasses inside it can narrow the age toward precision. Paleomagnetic insights have contributed greatly to the study of the travels of the continents, helping to show where they may have been with respect to one another. In the argot of geology, paleomagnetic specialists are sometimes called paleomagicians. Enough paleomagicians have been up and down the big roadcuts of the Palisades Sill to prepare what appears to be a Hilton for wrens and purple martins. Birds have shown no interest.

Near the end of the highway's groove in the sill, there opens a broad, forgettable view of the valley of the Hackensack. The road is descending toward the river. At an even greater angle, the sill—tilting westward—dives into the earth. Accordingly, as Karen Kleinspehn continues to move downhill she is going "upsection" through the diabase toward the top of the tilting sill. The texture of the rock becomes smoother, the crystals smaller, and soon she finds the contact where the magma—at 2000 degrees Fahrenheit—touched the country rock. The country rock was a shale, which had earlier been the deep muck of some Triassic lake, where the labyrinthodont amphibians lived, and paleoniscid fish. The diabase below the contact now is a smooth and uniform hard dark rock, no tweed—its crystals too small to be discernible, having had so little time to grow in the chill zone. The contact is a straight, clear line. She rests her hand across it. The heat of the magma penetrated about a hundred feet into the shale, enough to cook it, to metamorphose it, to turn it into spotted slate. Sampling the slate with her sledgehammer, she has to pound with even more persistence than before. "Some weird, wild minerals turn up in this stuff," she comments between swings. "The metamorphic aureole of this formation is about the hardest rock in New Jersey."

She moves a few hundred feet farther on, near the end of the series of cuts. Pin oaks, sycamores, aspens, cottonwoods have come in on the wind with milkweed and wisteria to seize living space between the rock and the road, although the environment appears to be less welcoming than the center of Carson Sink. There are fossil burrows in the slate—long stringers where Triassic animals travelled through the quiet mud, not far below the surface of the shallow lake. There is a huge rubber sandal by the road, a crate of broken eggs, three golf balls. Two are very cheap but one is an Acushnet Titleist. A soda can comes clinking down the interstate, moving ten miles an hour before the easterly winds of the traffic. The screen of trees damps the truck noise. Karen sits down to rest, to talk, with her back against a cottonwood. "Roadcuts can be a godsend. There's a series

of roadcuts near Pikeville, Kentucky—very big ones
—where you can see distributary channels in a river-
delta system, with natural levees, and with splay deposits going out from the levees into overbank deposits of shales and coal. It's a face-on view of the
fingers of a delta, coming at you—the Pocahontas
delta system, shed off the Appalachians in Missis-
sippian-Pennsylvanian time. You see river channels
that migrated back and forth across a valley and
were superposed vertically on one another through
time. You see it all there in one series of exposures,
instead of having to fit together many smaller pieces
of the puzzle."

Geologists on the whole are inconsistent drivers.
When a roadcut presents itself, they tend to lurch
and weave. To them, the roadcut is a portal, a fragment of a regional story, a proscenium arch that leads
their imaginations into the earth and through the
surrounding terrain. In the rock itself are the essential clues to the scenes in which the rock began to
form—a lake in Wyoming, about as large as Huron; a
shallow ocean reaching westward from Washington
Crossing; big rivers that rose in Nevada and fell
through California to the sea. Unfortunately, highway departments tend to obscure such scenes. They
scatter seed wherever they think it will grow. They
"hair everything over"—as geologists around the
country will typically complain.

"We think rocks are beautiful. Highway departments think rocks are obscene."

"In the North it's vetch."

"In the South it's the god-damned kudzu. You need a howitzer to blast through it. It covers the mountainsides, too."

"Almost all our stops on field trips are at roadcuts. In areas where structure is not well exposed, roadcuts are essential to do geology."

"Without some roadcuts, all you could do is drill a hole, or find natural streamcuts, which are few and far between."

"We as geologists are fortunate to live in a period of great road building."

"It's a way of sampling fresh rock. The road builders slice through indiscriminately, and no little rocks, no softer units are allowed to hide."

"A roadcut is to a geologist as a stethoscope is to a doctor."

"An X-ray to a dentist."

"The Rosetta Stone to an Egyptologist."

"A twenty-dollar bill to a hungry man."

"If I'm going to drive safely, I can't do geology."

In moist climates, where vegetation veils the earth, streamcuts are about the only natural places where geologists can see exposures of rock, and geologists have walked hundreds of thousands of miles in and beside streams. If roadcuts in the moist world are a kind of gift, they are equally so in other places. Rocks are not easy to read where natural outcrops are so deeply weathered that a hammer will virtually sink out of sight—for example, in piedmont Georgia. Make a fresh roadcut almost anywhere at all and geologists will close in swiftly, like missionaries racing anthropologists to a tribe just discovered up the Xingu.

"I studied roadcuts and outcrops as a kid, on long trips with my family," Karen says. "I was probably doomed to be a geologist from the beginning." She grew up in the Genesee Valley, and most of the long trips were down through Pennsylvania and the Virginias to see her father's parents, in North Carolina. On such a journey, it would have been difficult not to notice all the sheets of rock that had been bent, tortured, folded, faulted, crumpled—and to wonder how that happened, since the sheets of rock would have started out as flat as a pad of paper. "I am mainly interested in sedimentology, in sedimentary structures. It allows me to do a lot of field work. I'm not too interested in theories of what happens *x* kilometres down in the earth at certain temperatures and pressures. You seldom do field work if you're interested in the mantle. There's a little bit of the humanities that creeps into geology, and that's why I am in it. You can't prove things as rigorously as physicists or chemists do. There are no white coats in a

geology lab, although geology is going that way.
Under the Newark Basin are worn-down remains of
the Appalachians—below us here, and under that
valley, and so on over to the Border Fault. In the
West, for my thesis, I am working on a basin that
also formed on top of a preexisting deformed belt. I
can't say that the basin formed just like this one, but
what absorbs me are the mechanics of these succes-
sor basins, superposed on mountain belts. The Great
Valley in California is probably an example of a late-
stage compressional basin—formed as plates came
together. We think the Newark Basin is an extensional basin—formed as plates moved apart. In the
geologic record, how do we recognize the differences
between the two? I am trying to get the picture of
the basin as a whole, and what is the history that
you can read in these cuts. I can't synthesize all this
in one morning on a field trip, but I can look at the
rock here and then evaluate someone else's interpretation." She pauses. She looks back along the rockwall. "This interstate is like a knife wound all across
the country," she remarks. "Sure—you could do this
sort of thing from here to California. Anyone who
wants to, though, had better hurry. Before long, to
go all the way across by yourself will be a fossil experience. A person or two. One car. Coast to coast.
People do it now without thinking much about it.
Yet it's a most unusual kind of personal freedom—

particular to this time span, the one we happen to be in. It's an amazing, temporary phenomenon that will end. We have the best highway system in the world. It lets us do what people in no other country can do. And it is also an ecological disaster."

In June, every year, students and professors from Eastern colleges—with their hydrochloric-acid phials and their hammers and their Brunton compasses—head West. To be sure, there is plenty of absorbing geology under the shag of Eastern America, galvanic conundrums in Appalachian structure and intricate puzzles in history and stratigraphy. In no manner would one wish to mitigate the importance of the Eastern scene. Undeniably, though, the West is where the rocks are—"where it all hangs out," as someone in the United States Geological Survey has put it—and of Eastern geologists who do any kind of summer field work about seventy-five per cent go West. They carry state geological maps and the regional geological highway maps that have been published by the American Association of Petroleum Geologists—maps as prodigally colored as drip paintings and equally formless in their worm-trail-and-paramecium depictions of the country's uppermost rock. The maps give two dimensions but more than suggest the third. They tell the general age and story of the banks of the asphalt stream. Kleinspehn has been doing this for some years, getting into her

Minibago, old and overloaded, a two-door Ford,
heavy-duty springs, with odd pieces of the Rockies
under the front seat and a mountain tent in the gear
behind, to cross the Triassic lowlands and the Border
Fault and to rise into the Ridge and Valley Province,
the folded-and-faulted, deformed Appalachians—the
beginnings of a journey that above all else is physiographic, a journey that tends to mock the idea of a
nation, of a political state, as an unnatural subdivision of the globe, as a metaphor of the human ego
sketched on paper and framed in straight lines and in
riparian boundaries between unalterable coasts. The
United States: really a quartering of a continent, a
drawer in North America. Pull it out and prairie dogs
would spill off one side, alligators off the other—a
terrain crisscrossed with geological boundaries,
mammalian boundaries, amphibian boundaries: the
limits of the world of the river frog, the extent of the
Nugget Formation, the range of the mountain cougar. The range of the cougar is the cougar's natural
state, overlying segments of tens of thousands of
other states, a few of them proclaimed a nation. The
United States of America, with its capital city on the
Atlantic Coastal Plain. The change is generally dramatic as one province gives way to another; and
halfway across Pennsylvania, as you leave the quartzite ridges and carbonate valleys of the folded-and-
faulted mountains, you drop for a moment into

Cambrian rock near the base of a long climb, a ten-mile gradient upsection in time from the Cambrian into the Ordovician into the Silurian into the Devonian into the Mississippian (generally through the same chapters of the earth represented in the walls of the Grand Canyon) and finally out onto the Pennsylvanian itself, the upper deck, the capstone rock, of the Allegheny Plateau. Now even the Exxon map shows a new geology, roads running every which way like shatter lines in glass, following the crazed geometries of this deeply dissected country, whereas, before, the roads had no choice but to run northeast-southwest among the long ropy trends of the deformed mountains, following the endless ridges. On these transcontinental trips, Karen has driven as much as a thousand miles in a day at speeds that she has come to regard as dangerous and no less emphatically immoral. She has almost never slept under a roof, nor can she imagine why anyone on such a journey would want or need to; she "scopes out" her campsites in the late-failing light with strong affection for national forests and less for the three-dollar campgrounds where you roll out your Ensolite between two trailers, where gregarious trains honk like Buicks, and Yamahas on instruments climb escarpments in the night. The physiographic boundary is indistinct where you shade off the Allegheny Plateau and onto the stable craton, the continent's enduring core, its heartland, immemorially unstrained, the steady, predictable hedreocraton—the Stable Interior Craton. There are old mountains to the east, maturing mountains to the west, adolescent mountains beyond. The craton has participated on its edges in the violent creation of the mountains. But it remains intact within, and half a nation wide—the lasting, stolid craton, slowly, slowly downwasting. It has lost five centimetres since the birth of Christ. In much of Canada and parts of Minnesota and Wisconsin, the surface of the craton is Precambrian—earth-basement rock, the continental shield. Ohio, Indiana, Illinois, and so forth, the whole of what used to be called the Middle West, is shield rock covered with a sedimentary veneer that has never been metamorphosed, never been ground into tectonic hash—sandstones, siltstones, limestones, dolomites, flatter than the ground above them, the silent floors of departed oceans, of epicratonic seas. Iowa. Nebraska. Now with each westward township the country thickens, rises—a thousand, two thousand, five thousand feet—on crumbs shed off the Rockies and generously served to the craton. At last the Front Range comes to view—the chevroned mural of the mountains, sparkling white on gray, and on its outfanning sediments you are lifted into the Rockies and you plunge through a canyon to the Laramie Plains. "You go from one major geologic province to

another and—whoa!—you really know you're doing
it." There are mountains now behind you, mountains
before you, mountains that are set on top of mountains, a complex score of underthrust, upthrust, over-
thrust mountains, at the conclusion of which,
through another canyon, you come into the Basin
and Range. Brigham Young, when he came through
a neighboring canyon and saw rivers flowing out on
alluvial fans from the wall of the Wasatch to the flats
beyond, made a quick decision and said, "This is the
place." The scene suggested settling for it. The alternative was to press on beside a saline sea and then
across salt barrens so vast and flat that when micro-
wave relays would be set there they would not require towers. There are mountains, to be sure—off to
one side and the other: the Oquirrhs, the Stansburys,
the Promontories, the Silver Island Mountains. And
with Nevada these high, discrete, austere new ranges
begin to come in waves, range after range after north-
south range, consistently in rhythm with wide flat
valleys: basin, range; basin, range; a mile of height
between basin and range. Beside the Humboldt you
wind around the noses of the mountains, the Humboldt, framed in cottonwood—a sound, substantial,
year-round-flowing river, among the largest in the
world that fail to reach the sea. It sinks, it disappears,
in an evaporite plain, near the bottom of a series of
fault blocks that have broken out to form a kind of

stairway that you climb to go out of the Basin and
Range. On one step is Reno, and at the top is Donner
Summit of the uplifting Sierra Nevada, which has
gone above fourteen thousand feet but seems by no
means to have finished its invasion of the sky. The
Sierra is rising on its east side and is hinged on the
west, so the slope is long to the Sacramento Valley—
the physiographic province of the Great Valley—flat
and sea-level and utterly incongruous within its
flanking mountains. It was not eroded out in the
normal way of valleys. Mountains came up around it.
Across the fertile flatland, beyond the avocados,
stand the Coast Ranges, the ultimate province of the
present, the berm of the ocean—the Coast Ranges,
with their dry and straw-brown Spanish demeanor,
their shadows of the live oaks on the ground.

If you were to make that trip in the Triassic—
New York to San Francisco, Interstate 80, say
roughly at the end of Triassic time—you would
move west from the nonexistent Hudson River with
the Palisades Sill ten thousand feet down. The motions that will open the Atlantic are well under way
(as things appear in present theory), but the brine
has not yet come in. Behind you, in fact, where the
ocean will be, are several thousand miles of land—a
contiguous landmass, fragments of which will be
Africa, Antarctica, India, Australia. You cross the
Newark Basin. It is for the most part filled with red

mud. In the mud are tracks that seem to have been made by a two-ton newt. You come to a long, low, north-south-trending, black, steaming hill. It is a flow of lava that has come out over the mud and has cooled quickly in the air to form the dense smooth textures of basalt. Someday, towns and landmarks of this extruded hill will in one way or another take from it their names: Montclair, Mountainside, Great Notch, Glen Ridge. You top the rise, and now you can see across the rest of the basin to the Border Fault, and—where Whippany and Parsippany will be, some thirty miles west of New York—there is a mountain front perhaps seven thousand feet high. You climb this range and see more and more mountains beyond, and they are the folded-and-faulted Appalachians, but middle-aged and a little rough still at the edges, not caterpillar furry and worn-down smooth. Numbers do not seem to work well with regard to deep time. Any number above a couple of thousand years—fifty thousand, fifty million—will with nearly equal effect awe the imagination to the point of paralysis. This Triassic journey, anyway, is happening close to two hundred million years ago, or five per cent back into the existence of the earth. From the subalpine peaks of New Jersey, the descent is long and gradual to the lowlands of western Pennsylvania, where flat-lying sedimentary rocks begin to reach out across the craton—coals and sandstones, shales and limestones, slowly downwasting, Ohio, Indiana, Illinois, Iowa, erosionally losing an inch every thousand years. Where the Missouri will flow, past Council Bluffs, you come into a world of ruddy hills, Permian red, that continue to the far end of Nebraska, where you descend to the Wyoming flats. Sandy in places, silty, muddy, they run on and on, near sea level, all the way across Wyoming and into Utah. They are as red as brick. They will become the red cliffs and red canyons of Wyoming, the walls of Flaming Gorge. Triassic rock is not exclusively red, but much of it is red all over the world—red in the shales of New Jersey, red in the sandstones of Yunan, red in the banks of the Volga, red by the Solway Firth. Triassic redbeds, as they are called, are in the dry valleys of Antarctica, the red marls of Worcestershire, the hills of Alsace-Lorraine. The Petrified Forest. The Painted Desert. The South African redbeds of the Great Karroo. Triassic red rock is red through and through, and not merely weathered red on the surface, like the great Redwall Limestone of the Grand Canyon, which is actually gray. There may have been a superabundance of oxygen in the atmosphere from late Pennsylvanian through Permian and Triassic time. As sea level changed and changed again all through the Pennsylvanian, tremendous quantities of vegetation grew and then were drowned and buried, grew and then were drowned and buried—to become, eventually, seam upon seam of coal, interlayered with sandstones and shales. Living plants take in carbon dioxide, keep the carbon in their carbohydrates, and give up the oxygen to the atmosphere. Animals, from bacteria upward, then eat the plants and reoxidize the carbon. This cycle would go awry if a great many plants were buried. Their carbon would be buried with them—isolated in rock—and so the amount of oxygen in the atmosphere would build up. All over the world, so much carbon was buried in Pennsylvanian time that the oxygen pressure in the atmosphere quite possibly doubled. There is more speculation than hypothesis in this, but what could the oxygen do? Where could it go? After carbon, the one other thing it could oxidize in great quantity was iron— abundant, pale-green ferrous iron, which exists everywhere, in fully five per cent of crustal rock; and when ferrous iron takes on oxygen, it turns a ferric red. That may have been what happened—in time that followed the Pennsylvanian. Permian rock is generally red. Redbeds on an epic scale are the signs of the Triassic, when the earth in its rutilance may have outdone Mars.

As you come off the red flats to cross western Utah, nearly two hundred million years before the present, you travel in the dark, there being not one grain of evidence to suggest its Triassic appearance, no paleoenvironmental clue. Ahead, though, in eastern Nevada, is a line of mountains that are much of an age with the peaks of New Jersey—a little rounded, beginning to show age—and after you climb them and go down off their western slopes you discern before you the white summits of alpine fresh terrain, of new rough mountains rammed into thin air, with snow banners flying off the matterhorns, ridges, crests, and spurs. You are in central Nevada, about four hundred miles east of San Francisco, and after you have climbed these mountains you look out upon (as it appears in present theory) open sea. You drop swiftly to the coast, and then move on across moderately profound water full of pelagic squid, water that is quietly accumulating the sediments which—ages in the future—will become the roof rock of the rising Sierra. Tall volcanoes are standing in the sea. Then, at roughly the point where the Sierran foothills will end and the Great Valley will begin —at Auburn, California—you move beyond the shelf and over deep ocean. There are probably some islands out there somewhere, but fundamentally you are crossing above ocean crustal floor that reaches to the China Sea. Below you there is no hint of North America, no hint of the valley or the hills where Sacramento and San Francisco will be.

I used to sit in class and listen to the terms come floating down the room like paper airplanes. Geology was called a descriptive science, and with its pitted outwash plains and drowned rivers, its hanging tributaries and starved coastlines, it was nothing if not descriptive. It was a fountain of metaphor—of isostatic adjustments and degraded channels, of angular unconformities and shifting divides, of rootless mountains and bitter lakes. Streams eroded headward, digging from two sides into mountain or hill, avidly struggling toward each other until the divide between them broke down, and the two rivers that did the breaking now became confluent (one yielding to the other, giving up its direction of flow and going the opposite way) to become a single stream. Stream capture. In the Sierra Nevada, the

Yuba had captured the Bear. The Macho member of a formation in New Mexico was derived in large part from the solution and collapse of another formation. There was fatigued rock and incompetent rock and inequigranular fabric in rock. If you bent or folded rock, the inside of the curve was in a state of compression, the outside of the curve was under great tension, and somewhere in the middle was the surface of no strain. Thrust fault, reverse fault, normal fault—the two sides were active in every fault. The inclination of a slope on which boulders would stay put was the angle of repose. There seemed, indeed, to be more than a little of the humanities in this subject. Geologists communicated in English; and they could name things in a manner that sent shivers through the bones. They had roof pendants in their discordant batholiths, mosaic conglomerates in desert pavement. There was ultrabasic, deep-ocean, mottled green-and-black rock—or serpentine. There was the slip face of the barchan dune. In 1841, a paleontologist had decided that the big creatures of the Mesozoic were "fearfully great lizards," and had therefore named them dinosaurs. There were festooned crossbeds and limestone sinks, pillow lavas and petrified trees, incised meanders and defeated streams. There were dike swarms and slickensides, explosion pits, volcanic bombs. Pulsating glaciers. Hogbacks. Radiolarian ooze. There was almost enough resonance in some terms to stir the adolescent groin. The swelling up of mountains was described as an orogeny. Ontogeny, phylogeny, orogeny—accent syllable two. The Antler Orogeny, the Avalonian Orogeny, the Taconic, Acadian, Alleghenian Orogenies. The Laramide Orogeny. The center of the United States had had a dull geologic history—nothing much being accumulated, nothing much being eroded away. It was just sitting there conservatively. The East had once been radical—had been unstable, reformist, revolutionary, in the Paleozoic pulses of three or four orogenies. Now, for the last hundred and fifty million years, the East had been stable and conservative. The far-out stuff was in the Far West of the country—wild, weirdsma, a leather-jacket geology in mirrored shades, with its welded tuffs and Franciscan mélange (internally deformed, complex beyond analysis), its strike-slip faults and falling buildings, its boiling springs and fresh volcanics, its extensional disassembling of the earth.

There was, to be sure, another side of the page —full of geological language of the sort that would have attracted Gilbert and Sullivan. Rock that stayed put was called autochthonous, and if it had moved it was allochthonous. "Normal" meant "at right angles." "Normal" also meant a fault with a depressed hanging wall. There was a Green River Basin in Wyoming that was not to be confused with the Green River Basin in Wyoming. One was topographical and was *on* Wyoming. The other was structural and was *under* Wyoming. The Great Basin, which is centered in Utah and Nevada, was not to be confused with the Basin and Range, which is centered in Utah and Nevada. The Great Basin was topographical, and extraordinary in the world as a vastness of land that had no drainage to the sea. The Basin and Range was a realm of related mountains that coincided with the Great Basin, spilling over slightly to the north and considerably to the south. To anyone with a smoothly functioning bifocal mind, there was no lack of clarity about Iowa in the Pennsylvanian, Missouri in the Mississippian, Nevada in Nebraskan, Indiana in Illinoian, Vermont in Kansan, Texas in Wisconsinan time. Meteoric water, with study, turned out to be rain. It ran downhill in consequent, subsequent, obsequent, resequent, and not a few insequent streams.

As years went by, such verbal deposits would thicken. Someone developed enough effrontery to call a piece of our earth an epieugeosyncline. There were those who said interfluve when they meant between two streams, and a perfectly good word like mesopotamian would do. A cactolith, according to the American Geological Institute's *Glossary of Geology and Related Sciences*, was "a quasi-horizontal chonolith composed of anastomosing ductoliths, whose distal ends curl like a harpolith, thin like a sphenolith, or bulge discordantly like an akmolith or ethmolith." The same class of people who called one rock serpentine called another jacupirangite. Clinoptilolite, eclogite, migmatite, tincalconite, szaibelyite, pumpellyite. Meyerhofferite. The same class of people who called one rock paracelsian called another despujolsite. Metakirchheimerite, phlogopite, katzenbuckelite, mboziite, noselite, neighborite, samsonite, pigeonite, muskoxite, pabstite, aenigmatite. Joesmithite. With the X-ray diffractometer and the X-ray fluorescence spectrometer, which came into general use in geology laboratories in the late nineteen-fifties, and then with the electron probe (around 1970), geologists obtained ever closer examinations of the components of rock. What they had long seen through magnifying lenses as specimens held in the hand—or in thin slices under microscopes—did not always register identically in the eyes of these machines. Andesite, for example, had been given its name for being the predominant rock of the high mountains of South America. According to the machines, there is surprisingly little andesite in the Andes. The Sierra Nevada is renowned throughout the world for its relatively young and absolutely beautiful granite. There is precious little granite in the Sierra. Yosemite Falls, Half Dome, El Capitan— for the most part the "granite" of the Sierra is granodiorite. It has always been difficult enough to hold in the mind that a magma which hardens in the earth as granite will—if it should flow out upon the earth—harden as rhyolite, that what hardens within the earth as diorite will harden upon the earth as andesite, that what hardens within the earth as gabbro will harden upon the earth as basalt, the difference from pair to pair being a matter of chemical composition and the differences within each pair being a matter of texture and of crystalline form, with the darker rock at the gabbro end and the lighter rock the granite. All of that—not to mention such wee appendixes as the fact that diabase is a special texture of gabbro—was difficult enough for the layman to remember before the diffractometers and the spectrometers and the electron probes came along to present their multiplex cavils. What had previously been described as the granite of the world turned out to be a large family of rock that included granodiorite, monzonite, syenite, adamellite, trondh-jemite, alaskite, and a modest amount of true granite. A great deal of rhyolite, under scrutiny, became dacite, rhyodacite, quartz latite. Andesite was found to contain enough silica, potassium, sodium, and aluminum to be the fraternal twin of granodiorite. These points are pretty fine. The home terms still apply. The enthusiasm geologists show for adding new words to their conversation is, if anything, exceeded by their affection for the old. They are not about to drop granite. They say granodiorite when they are in church and granite the rest of the week.

When I was seventeen and staring up the skirts of Eastern valleys, I was taught the rudiments of what is now referred to as the Old Geology. The New Geology is the package phrase for the effects of the revolution that occurred in earth science in the nineteen-sixties, when geologists clambered onto sea-floor spreading, when people began to discuss continents in terms of their velocities, and when the interactions of some twenty parts of the globe became known as plate tectonics. There were few hints of all that when I was seventeen, and now, a shake later, middle-aged and fading, I wanted to learn some geology again, to feel the difference between the Old and the New, to sense if possible how the science had settled down a decade after its great upheaval, but less in megapictures than in day-to-day contact with country rock, seeing what had not changed as well as what had changed. The thought occurred to me that if you were to walk a series of roadcuts with a geologist something illuminating would in all likelihood occur. This was long before I met Karen Kleinspehn, or, for that matter, David Love, of the United States Geological Survey, or Anita Harris, also of the Survey, or Eldridge Moores, of the University of California at Davis, all of whom would eventually take me with them through various stretches of the continent. What I did first off was what anyone would do. I called my local geologist. I live in Princeton, New Jersey, and the man I got in touch with was Kenneth Deffeyes, a senior professor who teaches introductory geology at Princeton University. It is an assignment that is angled wide. Students who have little aptitude for the sciences are required to take a course or two in the sciences en route to some cerebral Valhalla dangled high by the designers of curriculum. Deffeyes' course is one that such students are drawn to select. He calls it ́Earth and Its Resources. They call it Rocks for Jocks.

Deffeyes is a big man with a tenured waistline. His hair flies behind him like Ludwig van Beethoven's. He lectures in sneakers. His voice is syllabic, elocutionary, operatic. He has been described by a colleague as "an intellectual roving shortstop, with more ideas per square metre than anyone else in the department—they just tumble out." His surname rhymes with "the maze." He has been a geological engineer, a chemical oceanographer, a sedimentary petrologist. As he lectures, his eyes search the hall. He is careful to be clear but also to bring forth the full promise of his topic, for he knows that while the odd jock and the pale poet are the white of his target the bull's-eye is the future geologist. Undergraduates do not come to Princeton intending to study geology.

When freshmen fill out cards stating their three principal interests, no one includes rocks. Those who will make the subject their field of major study become interested after they arrive. It is up to Deffeyes to interest them—and not a few of them—or his department goes into a subduction zone. So his eyes search the hall. People out of his course have been drafted by the Kansas City Kings and have set records in distance running. They have also become professors of geological geophysics at Caltech and of petrology at Harvard.

Deffeyes' own research has gone from Basin and Range sediments to the floor of the deep sea to unimaginable events in the mantle, but his enthusiasms are catholic and he appears to be less attached to any one part of the story than to the entire narrative of geology in its four-dimensional recapitulations of space and time. His goals as a teacher are ambitious to the point of irrationality: At the very least, he seems to expect a hundred mint geologists to emerge from his course—expects perhaps to turn on his television and see a certified igneous petrographer up front with the starting Kings. I came to know Deffeyes when I wondered how gold gets into mountains. I knew that most old-time hard-rock prospectors had little to go on but an association of gold with quartz. And I knew the erosional details of how gold comes out of mountains and into the rubble of streams. What I wanted to learn was what put the gold in the mountains in the first place. I asked a historical geologist and a geomorphologist. They both recommended Deffeyes. He explained that gold is not merely rare. It can be said to love itself. It is, with platinum, the noblest of the noble metals— those which resist combination with other elements. Gold wants to be free. In cool crust rock, it generally is free. At very high temperatures, however, it will go into compounds; and the gold that is among the magmatic fluids in certain pockets of interior earth may be combined, for example, with chlorine. Gold chloride is "modestly" soluble, and will dissolve in water that comes down and circulates in the magma. The water picks up many other elements, too: potassium, sodium, silicon. Heated, the solution rises into fissures in hard crust rock, where the cooling gold breaks away from the chlorine and—in specks, in flakes, in nuggets even larger than the eggs of geese —falls out of the water as metal. Silicon precipitates, too, filling up the fissures and enveloping the gold with veins of silicon dioxide, which is quartz.

When I asked Deffeyes what one might expect from a close inspection of roadcuts, he said they were windows into the world as it was in other times. We made plans to take samples of highway rock. I suggested going north up some new interstate to see what the blasting had disclosed. He said if you go north, in most places on this continent, the geology does not greatly vary. You should proceed in the direction of the continent itself. Go west. I had been thinking of a weekend trip to Whiteface Mountain, or something like it, but now, suddenly, a vaulting alternative came to mind. What about Interstate 80, I asked him. It goes the distance. How would it be? "Absorbing," he said. And he mused aloud: After 80 crosses the Border Fault, it pussyfoots along on morainal till that levelled up the fingers of the fold-belt hills. It does a similar dance with glacial debris in parts of Pennsylvania. It needs no assistance on the craton. It climbs a ramp to the Rockies and a fault-block staircase up the front of the Sierra. It is geologically shrewd. It was the route of animal migrations, and of human history that followed. It avoids melodrama, avoids the Grand Canyons, the Jackson Holes, the geologic operas of the country, but it would surely be a sound experience of the big picture, of the history, the construction, the components of the continent. And in all likelihood it would display in its roadcuts rock from every epoch and era.

In seasons that followed, I would go back and forth across the interstate like some sort of shuttle working out on a loom, accompanying geologists on purposes of their own or being accompanied by them from cut to cut and coast to coast. At

any location on earth, as the rock record goes down
into time and out into earlier geographies it touches
upon tens of hundreds of stories, wherein the face of
the earth often changed, changed utterly, and
changed again, like the face of a crackling fire. The
rock beside the road exposes one or two levels of the
column of time and generally implies what went on
immediately below and what occurred (or never occurred) above. To tell all the stories would be to tell
pretty much the whole of geology in many volumes
across a fifty-foot shelf, a task for which I am in
every conceivable way unqualified. I am a layman
who has travelled for a couple of years with a small
core sampling of academic and government geologists ranging in experience from a graduate student
to an authentic *éminence grise*, and what I intend to
do now is to distill the trips of those years. I wish to
make no attempt to speak for all geology or even to
sweep in a great many facts that came along. I want
to choose some things that interested me and through
them to suggest the general history of the continent
by describing events and landscapes that geologists
see written in rocks.

To poke around in a preliminary way, Deffeyes
and I went up to the Palisades Sill, where I was to
return with Karen Kleinspehn, borrowed some diabase with a ten-pound sledge, and then began to
travel westward, traversing the Hackensack Valley.

It was morning. Small airplanes engorged with businessmen were settling into Teterboro. Deffeyes
pointed out that if this were near the end of Wisconsinan time, when the ice was in retreat, those
airplanes would have been settling down through
several hundred feet of water, with the runway at
the bottom of a lake. Glacial Lake Hackensack was
the size of Lake Geneva and was host to many islands. It had the Palisades Sill for an eastern shoreline, and on the west the lava hill that is now known
as the First Watchung Mountain. The glacier had
stopped at Perth Amboy, leaving its moraine there to
block the foot of the lake, which the glacier fed with
meltwater as it retreated to the north. Nearly two
hundred million years earlier, the runway would
have been laid out on a baking red flat beside the
first, cooling Watchung—glowing from cracks, from
lava fountains, but generally black as carbon. Basalt
flows don't light up the sky. Three hundred million
years before that, the airplanes would have been settling down toward the same site through water—in
this instance, salt water—on the eastern shelf of a
broad low continent, where an almost pure limestone
was forming, because virtually nothing from the
worn-away continent was eroding into the shallow
sea. Three random moments from the upper ninth of
time.

In Paterson, I–80 chops the Watchung lava.

Walking the cut from end to end, Deffeyes picked up some peripheral shale—Triassic red shale. He put it in his mouth and chewed it. "If it's gritty it's a silt bed, and if it's creamy it's a shale," he said. "This is creamy. Try it." I would not have thought to put it in coffee. In the blocky basaltic wall of the road, there were many small pockets, caves the size of peas, caves the size of lemons. As magma approaches the surface of the earth, it is so perfused with gases that it fizzes like ginger ale. In cooling basalt, gas bubbles remain, and form these minicaves. For a century and more, nothing much fills them. Slowly, though, over a minimum of about a million years, they can fill with zeolite crystals. Until well after the Second World War, not a whole lot was known about the potential uses of zeolite crystals. Nor was it known where they could be found in abundance. Deffeyes did important early work in the field. His doctoral dissertation, which dealt with two basins and two ranges in Nevada, included an appendix that started the zeolite industry. Certain zeolites (there are about thirty kinds) have become the predominant catalysts in use in oil refineries, doing a job that is otherwise assigned to platinum. Now, in Paterson, Deffeyes searched the roadcut vugs (as the minute caves are actually called) looking for zeolites. Some vugs were large enough to suggest the holes that lobsters hide in. They did indeed contain a number of white fibrous zeolite crystals—smooth and soapy, of a type that resembled talc or asbestos—but the cut had been almost entirely cleaned out by professional and amateur collectors, undeterred by the lethal traffic not many inches away. Nearly all the vugs were now as empty as they had been in their first hundred years. In the shale beyond the lava we saw the burrows of Triassic creatures. An ambulance from Totowa flew by with its siren wailing.

We moved on a few miles into the Great Piece Meadows of the Passaic River Valley, flat as a lake floor, poorly drained land. A meadow in New Jersey is any wet spongy acreage where you don't sink in above your chin. Great Piece Meadows, Troy Meadows, Black Meadows, the Great Swamp—Whippany, Parsippany, Madison, and Morristown are strewn among the reeds. The whole region, very evidently, was the bottom of a lake, for a lake itself is by definition a sign of poor drainage, an aneurysm in a river, a highly temporary feature on the land. Some lakes dry up. Others disappear after the outlet stream, cutting back into the outlet, empties the water. This one—Glacial Lake Passaic—vanished about ten thousand years ago after the retreating glacier exposed what is now the Passaic Valley. The lake drained gradually into the new Passaic River, which fell a hundred feet into Glacial Lake Hackensack, and, en route, went over a waterfall that would one day in effect found the city of Paterson by turning its first mill wheel. At the time of its greatest extent, Lake Passaic was two hundred feet deep, thirty miles long, and ten miles wide, and seems to have been a scene of great beauty. Its margins are still decorated with sandspits and offshore bars, wave-cut cliffs and stream deltas, set in suburban towns. The lake's west shore was the worn-low escarpment of the Border Fault, and its most arresting feature was a hook-shaped basaltic peninsula that is now known to geologists as a part of the Third Watchung Lava Flow and to the people of New Jersey as Hook Mountain.

Deffeyes became excited as we approached Hook Mountain. The interstate had blasted into one toe of the former peninsula, exposing its interior to view. Deffeyes said, "Maybe someone will have left some zeolites here. I want them so bad I can taste them." He jumped the curb with his high-slung Geology Department vehicle, got out his hammers, and walked the cut. It was steep and competent, with brown oxides of iron over the felt-textured black basalt, and in it were tens of thousands of tiny vugs, a high percentage of them filled with pearl-lustred crystals of zeolite. To take a close look, he opened his hand lens—a small-diameter, ten-power Hastings triplet. "You can do a nice act in a jewelry store," he suggested. "You whip this thing out and you say the price is too high. These are beautiful crystals. Beautiful crystals imply slow growth. You don't get in a hurry and make something that nice." He picked up the sledge and pounded the cut, necessarily smashing many crystals as he broke their matrix free. "These crystals are like Vietnamese villages," he went on. "You have to destroy them in order to preserve them. They contain aluminum, silicon, calcium, sodium, and an incredible amount of imprisoned water. 'Zeolite' means 'the stone that boils.' If you take one small zeolite crystal, of scarcely more than a pinhead's diameter, and heat it until the water has come out, the crystal will have an internal surface area equivalent to a bedspread. Zeolites are often used to separate one kind of molecule from another. They can, for example, sort out molecules for detergents, choosing the ones that are biodegradable. They love water. In refrigerators, they are used to adsorb water that accidently gets into the Freon. They could be used in automobile gas tanks to adsorb water. A zeolite called clinoptilolite is the strongest adsorber of strontium and cesium from radioactive wastes. The clinoptilolite will adsorb a great deal of lethal material, which you can then store in a small space. When William Wyler made *The Big Country,* there was a climactic chase scene in which the bad guy was shot and came clattering down a canyon wall in what appeared to be a shower

of clinoptilolite. Geologists were on the phone to
Wyler at once. 'Loved your movie. Where was that
canyon?' There are a lot of zeolites in the Alps, in
Nova Scotia, and in North Table Mountain in Colorado. When I was at the School of Mines, I used to
go up to North Table Mountain just to wham
around. Some of the best zeolites in the world are in
this part of New Jersey."

There were oaks and maples on top of Hook
Mountain, and, in the wall of the roadcut, basal
rosettes of woolly mullein, growing in the rock. The
Romans drenched stalks of mullein with suet and
used them for funeral torches. American Indians
taught the early pioneers to use the long flannel
leaves of this plant as innersoles. Only three miles
west of us was the Border Fault, where the basin had
touched the range, where the stubby remnants of the
fault scarp are now under glacial debris. Deffeyes
said that the displacement along the fault—the even-
tual difference between two points that had been adjacent when the faulting began—exceeded fifteen
thousand feet. Of course, this happened over several
millions of years, and the mountains fronting the
basin were all the while eroding, so they were never
anything like fifteen thousand feet high. Generally,
though, in the late Triassic, there would have been
about a mile of difference, a mile of relief, between
basin and range. In flash floods, boulders came rain-

ing off the mountains and piled in fans at the edge of the basin, ultimately to be filled in with sands and muds and to form conglomerate, New Jersey's so-called Hammer Creek Conglomerate—multicircled, polka-dotted headcheese rock, sometimes known as puddingstone. Here where the basin met the range, the sediments piled up so much that after all of the erosion of two hundred million years what remains is three miles thick. "I was in a bar once in Austin, Nevada," Deffeyes said, "and there was a sudden torrential downpour. The bartender began nailing plywood over the door. I wondered why he was doing that, until boulders came tumbling down the main street of the town. When you start pulling a continent apart, you have a lot of consequences of the same event. Faulting produced this basin. Sediments filled it in. Pull things apart and you produce a surface vacancy, which is faulting, and a subsurface vacancy, which causes upwelling of hot mantle that intrudes as sills or comes out as lava flows. In the Old Geology, you might have seen a sill within the country rock and said, 'Ah, the sill came much later.' With the New Geology, you see that all this was happening more or less at one time. The continent was splitting apart and the ultimate event was the opening of the Atlantic. If you look at the foldbelt in northwest Africa, you see the other side of the New Jersey story. The folding there is of the same age as

the Appalachians, and the subsequent faulting is
Triassic. Put the two continents together on a map
and you will see what I mean. Fault blocks like this
one are still in evidence, but discontinuously, from
the Connecticut Valley to South Carolina. They are
all parts of the suite that opened the Atlantic seaway. The story is very similar in the Great Basin—in
the West, in the Basin and Range. The earth is splitting apart there, quite possibly opening a seaway. It
is not something that happened a couple of hundred
million years ago. It only began in the Miocene, and
it is going on today. What we are looking at here in
New Jersey is not just some little geologic feature,
like a zeolite crystal. This is the opening of the Atlantic. If you want to see happening right now what
happened here two hundred million years ago, you
can see it all in Nevada."

Basin. Fault. Range. Basin. Fault. Range. A mile of relief between basin and range. Stillwater Range. Pleasant Valley. Tobin Range. Jersey Valley. Sonoma Range. Pumpernickel Valley. Shoshone Range. Reese River Valley. Pequop Mountains. Steptoe Valley. Ondographic rhythms of the Basin and Range. We are maybe forty miles off the interstate, in the Pleasant Valley basin, looking up at the Tobin Range. At the nine-thousand-foot level, there is a stratum of cloud against the shoulders of the mountains, hanging like a ring of Saturn. The summit of Mt. Tobin stands clear, above the cloud. When we crossed the range, we came through a ranch on the ridgeline where sheep were fenced around a running brook and bales of hay were bright green. Junipers in the mountains were thickly hung with berries, and

the air was unadulterated gin. This country from
afar is synopsized and dismissed as "desert"—the
home of the coyote and the pocket mouse, the side-
blotched lizard and the vagrant shrew, the MX
rocket and the pallid bat. There are minks and river
otters in the Basin and Range. There are deer and
antelope, porcupines and cougars, pelicans, cor-
morants, and common loons. There are Bonaparte's
gulls and marbled godwits, American coots and Virginia rails. Pheasants. Grouse. Sandhill cranes. Fer-
ruginous hawks and flammulated owls. Snow geese.
This Nevada terrain is not corrugated, like the folded
Appalachians, like a tubal air mattress, like a rippled
potato chip. This is not—in that compressive manner
—a ridge-and-valley situation. Each range here is
like a warship standing on its own, and the Great
Basin is an ocean of loose sediment with these mountain ranges standing in it as if they were members of
a fleet without precedent, assembled at Guam to as-
sault Japan. Some of the ranges are forty miles long,
others a hundred, a hundred and fifty. They point
generally north. The basins that separate them—ten
and fifteen miles wide—will run on for fifty, a hundred, two hundred and fifty miles with lone, daisy-
petalled windmills standing over sage and wild rye.
Animals tend to be content with their home ranges
and not to venture out across the big dry valleys.
"Imagine a chipmunk hiking across one of these

basins," Deffeyes remarks. "The faunas in the high
ranges here are quite distinct from one to another.
Animals are isolated like Darwin's finches in the
Galápagos. These ranges are truly islands."

Supreme over all is silence. Discounting the cry
of the occasional bird, the wailing of a pack of coy-
otes, silence—a great spatial silence—is pure in the
Basin and Range. It is a soundless immensity with
mountains in it. You stand, as we do now, and look
up at a high mountain front, and turn your head and
look fifty miles down the valley, and there is utter
silence. It is the silence of the winter forests of the
Yukon, here carried high to the ridgelines of the
ranges. As the physicist Freeman Dyson has written
in *Disturbing the Universe*, "It is a soul-shattering
silence. You hold your breath and hear absolutely
nothing. No rustling of leaves in the wind, no rumbling of distant traffic, no chatter of birds or insects
or children. You are alone with God in that silence.
There in the white flat silence I began for the first
time to feel a slight sense of shame for what we were
proposing to do. Did we really intend to invade this
silence with our trucks and bulldozers and after a
few years leave it a radioactive junkyard?"

What Deffeyes finds pleasant here in Pleasant
Valley is the aromatic sage. Deffeyes grew up all over
the West, his father a petroleum engineer, and he
says without apparent irony that the smell of sage-

brush is one of two odors that will unfailingly bring upon him an attack of nostalgia, the other being the scent of an oil refinery. Flash floods have caused boulders the size of human heads to come tumbling off the range. With alluvial materials of finer size, they have piled up in fans at the edge of the basin. ("The cloudburst is the dominant sculptor here.") The fans are unconsolidated. In time to come, they will pile up to such enormous thicknesses that they will sink deep and be heated and compressed to form conglomerate. Erosion, which provides the material to build the fans, is tearing down the mountains even as they rise. Mountains are not somehow created whole and subsequently worn away. They wear down as they come up, and these mountains have been rising and eroding in fairly even ratio for millions of years—rising and shedding sediment steadily through time, always the same, never the same, like row upon row of fountains. In the southern part of the province, in the Mojave, the ranges have stopped rising and are gradually wearing away. The Shadow Mountains. The Dead Mountains, Old Dad Mountains, Cowhole Mountains, Bullion, Mule, and Chocolate Mountains. They are inselberge now, buried ever deeper in their own waste. For the most part, though, the ranges are rising, and there can be no doubt of it here, hundreds of miles north of the Mojave, for we are looking at a new seismic scar that runs as far as we can see. It runs along the foot of the mountains, along the fault where the basin meets the range. From out in the valley, it looks like a long, buff-painted, essentially horizontal stripe. Up close, it is a gap in the vegetation, where plants growing side by side were suddenly separated by several metres, where, one October evening, the basin and the range—Pleasant Valley, Tobin Range—moved, all in an instant, apart. They jumped sixteen feet. The erosion rate at which the mountains were coming down was an inch a century. So in the mountains' contest with erosion they gained in one moment about twenty thousand years. These mountains do not rise like bread. They sit still for a long time and build up tension, and then suddenly jump. Passively, they are eroded for millennia, and then they jump again. They have been doing this for about eight million years. This fault, which jumped in 1915, opened like a zipper far up the valley, and, exploding into the silence, tore along the mountain base for upward of twenty miles with a sound that suggested a runaway locomotive.

"This is the sort of place where you really do not put a nuclear plant," says Deffeyes. "There was other action in the neighborhood at the same time—in the Stillwater Range, the Sonoma Range, Pumpernickel Valley. Actually, this is not a particularly spectacular scarp. The lesson is that the whole thing

—the whole Basin and Range, or most of it—is alive. The earth is moving. The faults are moving. There are hot springs all over the province. There are young volcanic rocks. Fault scars everywhere. The world is splitting open and coming apart. You see a sudden break in the sage like this and it says to you that a fault is there and a fault block is coming up. This is a gorgeous, fresh, young, active fault scarp. It's growing. The range is lifting up. This Nevada topography is what you see *during* mountain building. There are no foothills. It is all too young. It is live country. This is the tectonic, active, spreading, mountain-building world. To a nongeologist, it's just ranges, ranges, ranges."

Most mountain ranges around the world are the result of compression, of segments of the earth's crust being brought together, bent, mashed, thrust and folded, squeezed up into the sky—the Himalaya, the Appalachians, the Alps, the Urals, the Andes. The ranges of the Basin and Range came up another way. The crust—in this region between the Rockies and the Sierra—is spreading out, being stretched, being thinned, being literally pulled to pieces. The sites of Reno and Salt Lake City, on opposite sides of the province, have moved apart fifty miles. The crust of the Great Basin has broken into blocks. The blocks are not, except for simplicity's sake, analogous to dominoes. They are irregular in shape. They more truly suggest stretch marks. Which they are. They trend north-south because the direction of the stretching is east-west. The breaks, or faults, between them are not vertical but dive into the earth at roughly sixty-degree angles, and this, from the outset, affected the centers of gravity of the great blocks in a way that caused them to tilt. Classically, the high edge of one touched the low edge of another and formed a kind of trough, or basin. The high edge —sculpted, eroded, serrated by weather—turned into mountains. The detritus of the mountains rolled into the basin. The basin filled with water—at first, it was fresh blue water—and accepted layer upon layer of sediment from the mountains, accumulating weight, and thus unbalancing the block even further. Its tilt became more pronounced. In the manner of a seesaw, the high, mountain side of the block went higher and the low, basin side went lower until the block as a whole reached a state of precarious and temporary truce with God, physics, and mechanical and chemical erosion, not to mention, far below, the agitated mantle, which was running a temperature hotter than normal, and was, almost surely, controlling the action. Basin and range. Integral fault blocks: low side the basin, high side the range. For five hundred miles they nudged one another across the province of the Basin and Range. With extra faulting, and whatnot, they took care of their own irregularities. Some had their high sides on the west, some on the east. The escarpment of the Wasatch Mountains—easternmost expression of this immense suite of mountains—faced west. The Sierra—the westernmost, the highest, the predominant range, with Donner Pass only halfway up it—presented its escarpment to the east. As the developing Sierra made its skyward climb—as it went on up past ten and twelve and fourteen thousand feet—it became so predominant that it cut off the incoming Pacific rain, cast a rain shadow (as the phenomenon is called) over lush, warm, Floridian and verdant Nevada. Cut it off and kept it dry.

We move on (we're in a pickup) into dusk— north up Pleasant Valley, with its single telephone line on sticks too skinny to qualify as poles. The big flanking ranges are in alpenglow. Into the cold clear sky come the ranking stars. Jackrabbits appear, and crisscross the road. We pass the darkening shapes of cattle. An eerie trail of vapor traverses the basin, sent up by a clear, hot stream. It is only a couple of feet wide, but it is running swiftly and has multiple sets of hot white rapids. In the source springs, there is a thumping sound of boiling and rage. Beside the springs are lucid green pools, rimmed with accumulated travertine, like the travertine walls of Lincoln Center, the travertine pools of Havasu Canyon, but these pools are too hot to touch. Fall in there and you are Brunswick stew. "This is a direct result of the crustal spreading," Deffeyes says. "It brings hot mantle up near the surface. There is probably a fracture here, through which the water is coming up to this row of springs. The water is rich in dissolved minerals. Hot springs like these are the source of vein-type ore deposits. It's the same story that I told you about the hydrothermal transport of gold. When rainwater gets down into hot rock, it brings up what it happens to find there—silver, tungsten, copper, gold. An ore-deposit map and a hot-springs map will look much the same. Seismic waves move slowly through hot rock. The hotter the rock, the slower the waves. Nowhere in the continental United States do seismic waves move more slowly than they do beneath the Basin and Range. So we're not woofing when we say there's hot mantle down there. We've measured the heat."

The basin-range fault blocks in a sense are floating on the mantle. In fact, the earth's crust everywhere in a sense is floating on the mantle. Add weight to the crust and it rides deeper, remove cargo and it rides higher, exactly like a vessel at a pier. Slowly disassemble the Rocky Mountains and carry the material in small fragments to the Mississippi Delta. The delta builds down. It presses ever deeper on the mantle. Its depth at the moment exceeds twenty-five thousand feet. The heat and the pressure are so great

down there that the silt is turning into siltstone, the
sand into sandstone, the mud into shale. For another
example, the last Pleistocene ice sheet loaded two
miles of ice onto Scotland, and that dunked Scotland
in the mantle. After the ice melted, Scotland came
up again, lifting its beaches high into the air. Isostatic adjustment. Let go a block of wood that you
hold underwater and it adjusts itself to the surface
isostatically. A frog sits on the wood. It goes down.
He vomits. It goes up a little. He jumps. It adjusts.
Wherever landscape is eroded away, what remains
will rise in adjustment. Older rock is lifted to view.
When, for whatever reason, crust becomes thicker, it
adjusts downward. All of this—with the central
image of the basin-range fault blocks floating in the
mantle—may suggest that the mantle is molten,
which it is not. The mantle is solid. Only in certain
pockets near the surface does it turn into magma and
squirt upward. The temperature of the mantle varies
widely, as would the temperature of anything that is
two thousand miles thick. Under the craton, it is described as chilled. By surface standards, though, it is
generally white hot, everywhere around the world—
white hot and solid but magisterially viscous, per-
mitting the crust above it to "float." Deffeyes was in
his bathtub one Saturday afternoon thinking about
the viscosity of the mantle. Suddenly he stood up
and reached for a towel. "Piano wire!" he said to

himself, and he dressed quickly and went to the library to look up a book on piano tuning and to calculate the viscosity of the wire. Just what he guessed—10^{22} poises. Piano wire. Look under the hood of a well-tuned Steinway and you are looking at strings that could float a small continent. They are rigid, but ever so slowly they will sag, will slacken, will deform and give way, with the exact viscosity of the earth's mantle. "And that," says Deffeyes, "is what keeps the piano tuner in business." More miles, and there appears ahead of us something like a Christmas tree alone in the night. It is Winnemucca, there being no other possibility. Neon looks good in Nevada. The tawdriness is refined out of it in so much wide black space. We drive on and on toward the glow of colors. It is still far away and it has not increased in size. We pass nothing. Deffeyes says, "On these roads, it's ten to the minus five that anyone will come along." The better part of an hour later, we come to the beginnings of the casino-flashing town. The news this year is that dollar slot machines are outdrawing nickel slot machines for the first time, ever.

D effeyes' purposes in coming to Nevada are pure and noble. His considerable energies appear to be about equally divided between the pursuit of pure science and the pursuit of noble metal. In order to enloft mankind's understanding of the basins, he has been taking paleomagnetic samples of basin sediments. He seeks insight into the way in which the rifting earth comes apart. He wants to perceive the subtle differences in the histories of one fault block and another. His ideas about silver, on the other hand, may send his children to college. This is, after all, Nevada, whose geology bought the tickets for the Spanish-American War. George Hearst found his fortune in the ground here. There were silver ores of such concentration that certain miners did nothing more to the heavy gray rocks than pack them up and ship them to Europe. To be sure, those days and those rocks—those supergene enrichments—are gone, but it has crossed the mind of Deffeyes that there may be something left for Deffeyes. Banqueting Sybarites surely did not lick their plates.

We rented the pickup in Salt Lake City—a white Ford. "If we had a bale of hay in here we'd be Nevada authentic," Deffeyes remarked, and he swept snow off the truck with a broom. November. Three inches on the ground and more falling, slanting in to us from the west. We squinted, and rubbed the insides of the windows, and passed low commercial buildings that drifted in and out of sight. WILD DUCKS & PHEASANTS PROCESSED. DEER CUT & WRAPPED. DRIVE-IN WINDOW. 7:00 TILL MIDNIGHT. Behind us we could not see, of course, the wall of the Wasatch, its triangles and pinnacles white, but westward of the city visibility improved, and soon other mountains— the Oquirrh Mountains—came looming out of the blankness, their strata steeply dipping and as distinct as the stripes of an awning. "Those are Pennsylvanian and Permian sandstones and limestones," Deffeyes said. "There was glaciation in the Southern Hemisphere at the time. The ice came and went. Sea level kept flapping up and down. So the deposition has a striped look."

When a mountain range comes up into the air, a

whole lot comes up with it. The event that had lifted the Oquirrhs—the stretching of the crust until it broke into blocks—was only among the latest of many episodes that have adjusted dramatically the appearance of central Utah. As we could plainly see from the interstate, the rock now residing in that striped mountainside had once been brutally shoved around—shoved, not pulled, and with such force that a large part of it had been tipped up more than ninety degrees, to and well beyond the vertical. Overturned. Such violence can happen on an epic scale. There is an entire nation in Europe that is upside down. It is not a superpower, but it is a whole country nonetheless—San Marino, overturned. Basin and Range faulting, on its own, has never overturned anything. The great fault blocks have a maximum tilt of thirty degrees. The event that so deformed the rock in the Oquirrhs took place roughly sixty million years ago—fifty-two million years before the Oquirrhs came into existence—and it was an event that made alpine fresh compressional mountains, which had their time here under the sun and were disassembled by erosion, taken down and washed away; and now those crazily upended stripes within the Oquirrhs are the evidence and fragmental remains of those ancestral mountains, brought up out of the earth and put on view as a component of new mountains. The new mountains—the mountains of

the Basin and Range—are packages variously containing rock that formed at one time or another during some five hundred and fifty million years, or an eighth of the earth's total time. It was thought until recently that older rock was in certain of the ranges, but improved techniques of dating have shown that not to be true. Seven-eighths of the earth's time is lost here, gone without evidence—rock that disintegrated and went off to be recycled. One-eighth, for all that, is no small amount of earth history, and as the great crustal blocks of the Basin and Range have tipped their mountains into the air, with individual faults offset as much as twenty thousand feet, they have brought to the surface and have randomly exposed former seafloors and basaltic dikes, entombed rivers and veins of gold, volcanic spewings and dunal sands—chaotic, concatenated shards of time. In the Basin and Range are the well-washed limestones of clear and sparkling shallow Devonian seas. There are dark, hard, cherty siltstones from some deep ocean trench full of rapidly accumulating Pennsylvanian guck. There are Triassic sediments rich in fossils, scattered pods of Cretaceous granite, Oligocene welded tuffs. There is not much layer-cake geology. The layers have too often been tortured by successive convulsive events.

The welded tuffs were the regional surface when basin-range faulting began. And for more than

twenty million previous years they had been the surface, the uppermost rock, with scant relief in the
topography of these vast volcanic plains, whose great
size and barren aspect are commensurate with the
magnitude of the holocaust that brought the rock
onto the land. Up through perhaps a hundred fissures, dikes, chimneys, vents, fractures came a vio-
lently expanding, exploding mixture of steam and
rhyolite glass, and, in enormous incandescent clouds,
heavier than air, it scudded across the landscape like
a dust storm. The volcanic ash that would someday
settle down on Herculaneum and Pompeii was a
light powder compared with this stuff, and as the
great ground-covering clouds oozed into the con-
tours of the existing landscape they sent streams hiss-
ing to extinction, and covered the streambeds and
then the valleys, and—with wave after wave of addi-
tional cloud—obliterated entire drainages like plas-
ter filling a mold. They filled in every gully and
gulch, cave, swale, and draw until almost nothing
stuck above a blazing level plain, and then more
clouds came exploding from below and in unim-
peded waves spread out across the plain. Needless to
say, every living creature in the region died. Single
outpourings settled upon areas the size of Massachusetts, and before the heavy ash stopped flowing it
had covered twenty times that. Moreover, it was hot
enough to weld. As the great clouds collapsed and

condensed, they formed a compact rock in large part consisting of volcanic glass. It was so thick—as much as three hundred metres thick—that crystals formed slowly in the cooling glass. "When you bury a countryside in that much rock so hot it welds, that is the ultimate environmental catastrophe," Deffeyes remarked. "I'm glad there hasn't been one recently."

The province, stung like that, sat still here for twenty-two million years, with volcanism continuing only on its periphery, while erosion worked on the tuff, making draws and gulches, modest valleys and unspectacular hills, but not extensively altering the essentially level plain. There was no repetition of the foaming, frothing outpourings that had completely changed many tens of thousands of square miles of the face of the earth, but so much disturbance arising from and within the underlying earth was obviously precursive of disturbances to follow, when the plains of welded tuff and some thousands of feet below them began to rift into crustal blocks and become the Basin and Range.

The basins filled immediately with water, and life came into the lakes. "Late-Miocene fossils are the earliest we have wherever we have found fossils in those lakebeds," Deffeyes said. "So Basin and Range faulting can be dated to the late Miocene—about eight million years ago." Gradually, as the rain shadow lengthened, the lakes "turned chemical"—

became saline or alkaline (bitter)—and eventually they dried up. There are basalt flows in the Basin and Range that are also post-Miocene—lavas that poured out on the surface well after the block faulting had begun, like the Watchungs of New Jersey. There are ruins of cinder cones—evidence of fairly recent local violence—and, in the basins and on the ranges, widespread falls of light ash from volcanoes beyond the province. You see, too, the stream deltas, shoreline terraces, and wave-cut cliffs of big lakes that came into the Great Basin after Pleistocene glaciation began. The change in world climate that made ice in the north temporarily preempted the rain shadow and dropped into the Great Basin torrents from the sky. In a region where evaporation had greatly exceeded precipitation, the reverse was now the case, and the big lakes in time connected the basins and made islands of the ranges—Lake Manlius (its bed is now, in part, Death Valley); Lake Lahontan, near Reno (its bed is now, in part, the Humboldt and Carson sinks); and Lake Bonneville. Lake Bonneville grew until it was the size of Lake Erie. Then it grew some more. At Red Rock Pass, in Idaho, it spilled over the brim of the Great Basin and into the Snake River plain. By now it was as large as Lake Michigan. It was not a glacial lake, just a sort of side effect of the distant glaciation, and it sat there for thousands of years with limestone terraces forming and waves cutting benches at the shoreline. Eventually, it began to drop, in stages, pausing wherever evaporation and precipitation were in temporary equilibrium, and more benches were cut and more terraces were made, and then as the rain shadow took over again, the water shrank back past Erie size and kept on shrinking and turning more and more chemical and getting smaller and shallower and shallower and smaller and near the end of its days became the Great Salt Lake.

The Great Salt Lake reached out to our right and disappeared in snow. In a sense, there was no beach. The basin flatness just ran to the lake and kept on going, wet. The angle formed at the shoreline appeared to be about 179.9 degrees. There were dark shapes of islands, firmaments in the swirling snow—elongate, north-south-trending islands, the engulfed summits of buried ranges. "Chemically, this is one of the toughest environments in the world," Deffeyes said. "You swing from the saltiest to the most dilute waters on the planet in a matter of hours. Some of the most primitive things living are all that can take that. The brine is nearly saturated with sodium chloride. For a short period each year, so much water comes down out of the Wasatch that large parts of the lake surface are relatively fresh. Any creature living there gets an osmotic shock that amounts to hundreds of pounds per square inch. No

higher plants can take that, no higher animals—no
multicelled organisms. Few bacteria. Few algae.
Brine shrimp, which do live there, die by the millions
from the shock."

I have seen the salt lake incredibly beautiful in
winter dusk under snow-streamer curtains of cloud
moving fast through the sky, with the wall of the
Wasatch a deep rose and the lake islands rising from
what seemed to be rippled slate. All of that was now
implied by the mysterious shapes in the foreshorten-
ing snow. I didn't mind the snow. One June day,
moreover, with Karen Kleinspehn—on her way west
for summer field work—I stopped in the Wasatch for
a picnic of fruit and cheese beside a clear Pyrenean
stream rushing white over cobbles of quartzite and
sandstone through a green upland meadow—cattle
in the meadow, cottonwoods along the banks of this
clear, fresh, suggestively confident, vitally ignorant
river, talking so profusely on its way to its fate,
which was to move among paradisal mountain landscapes until, through a terminal canyon, the Great
Basin drew it in. No outlet. Three such rivers feed
the Great Salt Lake. It does indeed consume them.
Descending, we ourselves went through a canyon so
narrow that the Union Pacific Railroad was in the
median of the interstate and on into an even steeper
canyon laid out as if for skiing in a hypnotizing
rhythm of christiania turns under high walls of rose-

brick Nugget sandstone and brittle shattered marine limestone covered with scrub oaks. "Good God, we are dropping out of the sky," said Kleinspehn, hands on the wheel, plunging through the big sheer road-cuts, one of which suddenly opened to distance, presented the Basin and Range.

" 'This is the place.' "

"You can imagine how he felt."

In the foreground was the alabaster city, with its expensive neighborhoods strung out along the Wasatch Fault, getting ready to jump fifteen feet. In the distance were the Oquirrhs, the Stansburys, the lake. Sunday afternoon and the Mormons were out on the flats by the water in folding chairs at collapsible tables, end to end like refectory tables, twenty people down to dinner, with acres of beach-flat all to themselves and seagulls around them like sacred cows. To go swimming, we had to walk first— several hundred yards straight out, until the water was ankle-deep. Then we lay down on our backs and floated. I have never been able to float. When I took the Red Cross tests, age nine to fifteen, my feet went down and I hung in the water with my chin wrenched up like something off Owl Creek Bridge. I kicked, slyly kicked to push my mouth above the surface and breathe. I could not truly float. Now I tried a backstroke and, like some sort of hydrofoil, went a couple of thousand feet on out over the lake.

Only my heels, rump, and shoulder blades seemed to be wet. I rolled over and crawled. I could all but crawl on my hands and knees. And this was June, at the south end—the least salty season, the least salty place in the whole of the Great Salt Lake.

Rolling up on one side, and propped on an elbow, I could see the Promontory Mountains across the water to the north, an apparent island but actually a peninsula, reaching southward into the lake. In 1869, a golden spike was carried into the Promontories and driven into a tie there to symbolize the completion of the first railroad to cross the North American continent—exactly one century before the first footprint on the moon, a span of time during which Salt Lake City and Reno would move apart by one human stride. In that time, also, the railroad twice became dissatisfied with the local arrangements of its roadbed—losing affection for the way of the golden spike (over the mountains) and building a causeway and wooden trestle across the lake itself, barely touching the Promontory peninsula at its southern tip. In the late nineteen-fifties, the trestle section was replaced by rock. The causeway traverses the lake like a solid breakwater, dividing it into halves. The principal rivers that flow into the Great Salt Lake all feed the southern half. The water on the north side of the causeway is generally a foot or two lower and considerably saltier than the water on the other side. Evaporate one cupful of Great Salt Lake North and you have upward of a third of a cup of salt. Evaporate a cupful of Great Salt Lake South and you have about a quarter of a cup of salt, or—nonetheless—eight times as much as from a cup of the ocean. As the lake drew at our bodies, trying to pull fresh water through our skins, it closed our pores tight and our lips swelled and became slightly numb. The water stung savagely at the slightest scratch and felt bitter as strep in the back of the throat.

We filled a bag with eggstones from the bottom, with oolites, the Salt Lake sand. It was by no means ordinary sand—not the small, smoothed-off ruins of mountains, carried down and dumped by rivers. It was sand that had formed in the lake. Just as rain-drops are created around motes of dust, oolites form around bits of rock so tiny that in wave-tossed water they will stir up and move. They move, and settle, move, and settle. And while they are up in the water calcium carbonate forms around them in layer after layer, building something like a pearl. Slice one in half with a diamond saw and you reveal a perfect bull's-eye, or, as its namer obviously imagined it, a stone egg, white and yolk—an oolite. Underwater on the Bahama Banks are sweeping oolitic dunes. When a geologist finds oolites embedded in rock—in, say, some Cambrian outcrop in the Lehigh Valley—the

Bahamas come to mind, and the Great Salt Lake, and, by inference, a shallow, lime-rich Cambrian sea. Our sample bag was like a ten-pound sack of sugar. I rolled over on my back, set it on my stomach, and, floating a little lower, kicked in to shore.

On the firm flat beach of the Great Salt Lake were many hundreds of thousands of brine flies— broad dark patches of them hopping and buzzing a steady collective electrical hum. A sacred gull made short bursts through the brine flies, its bill clapping. Three years before gulls ate crickets and saved the Mormons, Kit Carson shot gulls to feed the starving emigrants. Gulls, though, and brine flies are natural survivors. Now, at the end of spring runoff, dead creatures were everywhere. Osmotic shock had killed shrimp outnumbering the flies. Corpses, a couple of centimetres each, lay in hydrogen-sulphide decaying stink. Interlayered with the oolites on the bottom of the lake was a kind of galantine of brine shrimp, the greasy black muck of quintillions dead.

Salt crystals clung like snow to our hair, and were spread on our faces like powder. In man-made ponds near the shore, the sun was making Morton's salt. Spaced along the beach were water towers, courtesy of the State of Utah. You pulled a rope and took a shower.

And now in the autumn snow, Deffeyes and I could see shoreline terraces of Lake Bonneville a thousand feet above us on mountain slopes. That a lake so deep had been brought down to a present average depth of thirteen feet was food for melancholia. Still shrinking, it had long since become the world's second-deadest body of water. In a couple of hundred years, it could match the Dead Sea.

"Mother of God, that's nice," said Deffeyes suddenly, braking down the pickup on the shoulder of the road. The tip of the nose of the Stansbury Mountains had been sliced off by the interstate to reveal a sheer and massive section of handsome blue rock, thinly bedded, evenly bedded, forty metres high. Its parallel planes were tilting, dipping, gently to the east, with the exception of some confused and crumpled material that suggested a snowball splatted against glass, or a broken-down doorway in an otherwise undamaged wall. Deffeyes said, "Let's Richter the situation," and he got out and crossed the road. With his hammer, he chipped at the rock, puzzled the cut. He scraped the rock and dropped acid on the scrapings. Tilted by the western breeze, the snow was dipping sixty degrees east. The bedding planes were dipping twenty degrees east; and the stripes of Deffeyes' knitted cap were dipping fifty degrees north. The cap had a big tassel, and with his gray-wisped hair coming out from under in a curly mélange he looked like an exaggerated elf. He said he thought he knew what had caused "that big goober" in the rock, and it was almost certainly not a manifestation of some major tectonic event—merely local violence, a cashier shot in a grab raid, an item for an inside page. The cut was mainly limestone, which had collected as lime mud in an Ordovician sea. The goober was dolomite.

Limestone is calcium carbonate. Dolomite is calcium carbonate with magnesium added. Together they are known as the carbonate rocks. Deffeyes was taught in college that while it seemed obvious to infer that magnesium precipitating out of water changes limestone into dolomite there was no way to check this out empirically because dolomite was forming nowhere in the world. Deffeyes found that impossible to believe. Deffeyes was already a uniformitarian—a geologist who believes that the present is the key to the past, that if you want to understand how a rock is formed you go watch it forming now. Watch basalt flows at Mauna Loa. Watch the festooned crossbeddings of future sandstones being sketched by the currents of Hatteras. Watch a flooding river blanket the tracks of a bear. Surely, somewhere, he thought, limestone must be changing into dolomite now. Not long after graduate school, he and two others went to Bonaire, in the Netherlands Antilles, where they found a lagoon that was concentrating under the sun and "making a juice very rich in magnesium." The juice was flowing through the limestone below and changing it into dolomite. They presented the news in *Science*. When the rock of this big Utah roadcut had been the limy bottom of the Ordovician sea, the water had been so shallow that the lime mud had occasionally been above the surface and had dried out and cracked into chips, and then the water rose and the chips became embedded in more lime mud, and the process happened again and again so that the limestone now is a self-containing breccia studded with imprisoned chips—an accident so lovely to the eye you want to slice the rock and frame it.

In age, the blue stone approached five hundred million years. Captain Howard Stansbury, USA, whose name would rest upon the mountains of which the rock was a component, was approaching fifty when he came into the Great Basin in 1849. He had been making lighthouses in Florida. The government preferred that he survey the salt lake. With sixteen mules, a water keg, and some India-rubber bags, he circumambulated the lake, and then some. People told him not to try it. He ran out of water but not of luck. And he came back with a story of having seen —far out on the westward flats—scattered books, clothing, trunks, tools, chains, yokes, dead oxen, and abandoned wagons. The Donner Party went around the nose of the Stansburys in late August, 1846, rock on their left, lake marshes on their right. This huge blue roadcut, in its supranatural way, would have frightened them to death. They must have filed along just about where Deffeyes had parked the pickup, on the outside shoulder of the interstate. Deffeyes and I went back across the road, waiting first for a three-unit seven-axle tractor-trailer to pass. Deffeyes described it as "a freaking train."

Stansbury Mountains, Skull Valley . . . The Donner Party found good grass in Skull Valley, and good water, and a note by a post at a spring. It had been torn to shreds by birds. The emigrants pieced it together. "Two days—two nights—hard driving—cross desert—reach water." They went out of Skull Valley over the Cedar Mountains into Ripple Valley and over Grayback Mountain to the Great Salt Lake Desert. Grayback Mountain was basalt, like the Watchungs of New Jersey. The New Jersey basalt flowed about two hundred million years ago. The Grayback Mountain basalt flowed thirty-eight million years ago. Well into this century, it was possible to find among the dark-gray outcrops of Grayback Mountain pieces of wagons and of oxhorn, discarded earthenware jugs. The snow suddenly gone now, and in cold sunshine, Deffeyes and I passed Grayback Mountain and then had the Great Salt Lake Desert before us—the dry bed of Bonneville—broader than the periphery of vision. The interstate runs close to but not parallel to the wagon trail, which trends a little more northwesterly. The wagon trail aims directly at Pilot Peak of the Pilot Range, which we could see clearly, upward of fifty miles away—a pyramidal summit with cloud coming off it in the wind like a banner unfurling. Across the dry lakebed, the emigrants homed on Pilot Peak, standing in what is now Nevada, above ten thousand feet. Along the fault scarp, at the base of Pilot Peak, are cold springs. When the emigrants arrived at the springs, their tongues were bloody and black.

"Imagine those poor sons of bitches out here with their animals, getting thirsty," Deffeyes said. "It's a wonder they didn't string the guy that invented this route up by his thumbs."

The flats for the most part were alkaline, a leather-colored mud superficially dry. Dig down two inches and it was damp and greasy. Come a little rain and an ox could go in to its knees. The emigrants made no intended stops on the Great Salt Lake Desert. They drove day and night for the Pilot Range. In the day, they saw mirages—towers and towns and shimmering lakes. Sometimes the lakes were real—playa lakes, temporary waters after a storm. Under a wind, playa lakes move like puddles of mercury in motion on a floor—two or three hundred square miles of water on the move, here today, there tomorrow, gone before long like a mirage, leaving wagons mired in unimagined mud. Very few emigrants chose to cross the Bonneville flats, although the route was promoted as a shortcut—"a nigher route"—rejoining the main migration four basins into Nevada. It was the invention of Lansford Hastings and was known as the Hastings Cutoff. Hastings wrote the helpful note in Skull Valley. His route was geologically unfavorable, but this escaped his knowledge and notice. His preoccupations were with politics. He wished to become President of California. He saw California—for the moment undefendably Mexican—as a new nation, under God, conceived at liberty and dedicated to the proposition that anything can be accomplished through promotion: President Lansford Hastings, in residence in a Western White House. His strategy for achieving high office was to create a new shortcut on the way west, to promote both the route and the destination through recruiting and pamphleteering, to attract emigrants by the thousands year after year, and as their counsellor and deliverer to use them as constituent soldiers in the promised heaven. He camped beside the trail farther east. He attracted the Donners. He attracted Reeds, Kesebergs, Murphys, McCutchens, drew them southward away from the main trek and into the detentive scrub oak made fertile by the limestones of the Wasatch. The Donners were straight off the craton—solid and trusting, from Springfield, Illinois. Weeks were used hacking a path

through the scrub oaks, which were living barbed
wire. Equipment was abandoned on the Bonneville
flats to lighten up loads in the race against thirst.
Even in miles, the nigher route proved longer than
the one it was shortcutting, on the way to a sierra
that was named for snow.

Deffeyes and I passed graffiti on the Bonneville
flats. There being nothing to carve in and no medium
substantial enough for sprayed paint, the graffitists
had lugged cobbles out onto the hard mud—stones
as big as grapefruit, ballast from the interstate—and
in large dotted letters had written their names: ROSS,
DAWN, DON, JUDY, MARK, MOON, ERIC, fifty or sixty
miles of names. YARD SALE. Eric's lithography was in
basalt and dolomite, pieces of Grayback Mountain,
apparently, pieces of the Stansburys. His name, if it
sits there a century or so, will eventually explode.
Salt will work into the stones along the grain
boundaries. When this happens, water evaporates
out of the salt, and salt crystals keep collecting and
expanding until they explode the rock. In Death Valley are thousands of little heaps of crumbs that were
once granite boulders. Salt exploded them. Salt gets
into fence posts and explodes them at the base.

Near the far side of Utah, the flats turned blind-
ing white, corn-snow white, and revolving winds
were making devils out of salt. Over the whiteness
you could see the salt go off the curve of the earth.

When the drivers of jet cars move at Mach .8 over
the Bonneville Salt Flats, they feel that they are always about to crest a hill. Dig into the salt and it
turns out to be a crusty white veneer, like cake icing,
more than an inch thick—an almost pure sodium
chloride. Below it are a few inches of sand-size salt
particles, and below them a sort of creamy yogurt
mud that is the color of blond coffee. In much the
manner in which these salts were left behind by the
shrinking outline of the saline lake, there were times
around the edges of North America when the shrinking ocean stranded bays that gradually dried up and
left plains of salt. When the ocean came back, came
up again, it spread inland over the salt, which was
not so much dissolved as buried, under layers of
sediment washing in from the continent. With the
weight of more and more sediment, the layers of salt
went deep. Salt has a low specific gravity and is
very plastic. Pile eight thousand feet of sediment on
it and it starts to move. Slowly, blobularly, it collects
itself and moves. It shoves apart layers of rock. It
mounds upon itself, and, breaking its way upward,
rises in mushroom shape—a salt dome. Still rising
into more shales and sandstones, it bends them into
graceful arches and then bursts through them like a
bullet shooting upward through a splintering floor.
The shape becomes a reverse teardrop. Generally,
after the breakthrough, there will be some big layers

of sandstone leaning on the salt dome like boards leaning up against a wall. The sandstone is permeable and probably has a layer of shale above it, which is not permeable. Any fluid in the sandstone will not only be trapped under the shale but will also be trapped by the impermeable salt. Enter the strange companionship of oil and salt. Oil also moves after it forms. You never find it where God put it. It moves great distances through permeable rock. Unless something traps it, it will move on upward until it reaches daylight and turns into tar. You don't run a limousine on tar, let alone a military-industrial complex. If, however, the oil moves upward through inclined sandstone and then hits a wall of salt, it stops, and stays—trapped. Run a little drill down the side of a salt dome and when you hit "sand" it may be full of oil. In the Gulf of Mexico were many of the bays that dried up covered with salt. Where the domes are now, there are towers in the Gulf. A number of salt domes are embedded in the Mississippi Delta, and have been mined. There are rooms inside them with ceilings a hundred feet high—room after room after room, like convention halls, with walls, floors, and ceilings of salt, above ninety-nine per cent pure.

Deffeyes was saying, "It's likely that in under this salt flat are mountain structures just as complicated as any of the ranges. They're just buried."

We picked up some shattered limestones and

welded tuffs close by the Nevada state line. The tuff was hard, heavy, crystalline rock, freckled with feldspars and quartz. You would never dig a city out of that. The ranges now were anything but buried, and Pilot Peak reached above the shadowed basin and high into sunlight, a mile above its valleys. Soon we were climbing the Toano Range. "Here comes another roadcut," said Deffeyes near the summit. "You can feel them coming on. The Taconic Parkway would drive you nuts. I–80 gives you one when you're ready for it." What it gave in the Toanos was granite—not some sibling, son, or cousin but granite himself: sparkling black hornblendes evenly spaced through a snowy field of feldspars and quartz. It was of much the same age as the celebrated rock of the Sierra. Its presence here suggested that the great crustal meltings in the tectonic drama farther west put out enough heat even in eastern Nevada to cook up this batch of fresh granite.

In this manner we moved along from roadcut to roadcut, range to range, like barnyard poultry pecking up rock, seeing what the fault blocks had lifted from below. We crossed the Goshute Valley and went up into the Pequops into red Devonian shales, Devonian siltstones, Devonian limestones—a great many millions of years older than the granite, and from another world. These were marine rocks (by and large), full of crinoids and other marine fossils.

Nothing about their appearance differed from sediment that might have collected over Illinois or Iowa in midcontinental, epicratonic seas. They provided not so much as a hint that they were actually from the continental shelf, that Pequop Summit is more or less where North America ended in Devonian time. The first attempt to move covered wagons directly across the continent to California ended at the Pequops, too. The wagons were abandoned at a spring by the eastern base of the mountains, a short hike off the interstate. Later emigrants made cooking fires with the wood of the wagons. Deffeyes was spitting out the siltstones but chewing happily on the shales.

The oolites of the Great Salt Lake were forming in the present. The dolomite of the Stansbury Mountains was five hundred million years old. The tuff had been welded for thirty million years. The age of the granite was a hundred million years. The rock of Pequop Summit was four times as old as that. On a scale of zero to five hundred, those samplings were bunched toward the extremes, with nothing representing the middle three hundred million years. That was just chance, though—just what the faults had happened to throw up—and farther down the road, at Golconda, would come a full-dress two-hundred-million-year-old Triassic show.

Geologists mention at times something they call

the Picture. In an absolutely unidiomatic way, they have often said to me, "You don't get the Picture." The oolites and dolomite—tuff and granite, the Pequop siltstones and shales—are pieces of the Picture. The stories that go with them—the creatures and the chemistry, the motions of the crust, the paleoenvironmental scenes—may well, as stories, stand on their own, but all are fragments of the Picture.

The foremost problem with the Picture is that ninety-nine per cent of it is missing—melted or dissolved, torn down, washed away, broken to bits, to become something else in the Picture. The geologist discovers lingering remains, and connects them with dotted lines. The Picture is enhanced by filling in the lines—in many instances with stratigraphy: the rock types and ages of strata, the scenes at the times of deposition. The lines themselves to geologists represent structure—folds, faults, flat-lying planes. Ultimately, they will infer why, how, and when a structure came to be—for example, why, how, and when certain strata were folded—and that they call tectonics. Stratigraphy, structure, tectonics. "First you read ze Kafka," I overheard someone say once in a library elevator. "Ond zen you read ze Turgenev. Ond zen ond only zen are—you—ready—for—ze Tolstoy."

And when you have memorized Tolstoy, you

may be ready to take on the Picture. Multidimensional, worldwide in scope and in motion through time, it is sometimes called the Big Picture. The Megapicture. You are cautioned not to worry if at first you do not wholly see it. Geologists don't see it, either. Not all of it. The modest ones will sometimes scuff a boot and describe themselves and their colleagues as scientific versions of the characters in John Godfrey Saxe's version of the Hindu fable of the blind men and the elephant. "We are blind men feeling the elephant," David Love, of the Geological Survey, has said to me at least fifty times. It is not unknown for a geological textbook to include snatches of the poem.

> *It was six men of Indostan*
> *To learning much inclined,*
> *Who went to see the Elephant*
> *(Though all of them were blind).*
> *That each by observation*
> *Might satisfy his mind.*

The first man of Indostan touches the animal's side and thinks it must be some sort of living wall. The second touches a tusk and thinks an elephant is like a spear. The others, in turn, touch the trunk, an ear, the tail, a knee—"snake," "fan," "rope," "tree."

> *And so these men of Indostan*
> *Disputed loud and long,*

Each in his own opinion
Exceeding stiff and strong,
Though each was partly in the right,
And all were in the wrong!

The blind men and the elephant are kept close
at hand mainly to slow down what some graduate
students refer to as "arm-waving"—the delivery,
with pumping elbows, of hypotheses so breathtak-
ingly original that the science seems for the moment
more imaginative than descriptive. Where it is solid,
it is imaginative enough. Geologists are famous for
picking up two or three bones and sketching an entire and previously unheard-of creature into a landscape long established in the Picture. They look at
mud and see mountains, in mountains oceans, in
oceans mountains to be. They go up to some rock
and figure out a story, another rock, another story,
and as the stories compile through time they connect
—and long case histories are constructed and written
from interpreted patterns of clues. This is detective
work on a scale unimaginable to most detectives,
with the notable exception of Sherlock Holmes, who
was, with his discoveries and interpretations of little
bits of grit from Blackheath or Hampstead, the first
forensic geologist, acknowledged as such by geologists to this day. Holmes was a fiction, but he started
a branch of a science; and the science, with careful

inference, carries fact beyond the competence of invention. Geologists, in their all but closed conversation, inhabit scenes that no one ever saw, scenes of global sweep, gone and gone again, including seas, mountains, rivers, forests, and archipelagoes of aching beauty rising in volcanic violence to settle down quietly and then forever disappear—*almost* disappear. If some fragment has remained in the crust somewhere and something has lifted the fragment to view, the geologist in his tweed cap goes out with his hammer and his sandwich, his magnifying glass and his imagination, and rebuilds the archipelago.

I once dreamed about a great fire that broke out at night at Nasser Aftab's House of Carpets. In Aftab's showroom under the queen-post trusses were layer upon layer and pile after pile of shags and broadlooms, hooks and throws, para-Persians and polyesters. The intense and shrivelling heat consumed or melted most of what was there. The roof gave way. It was a night of cyclonic winds, stabs of unseasonal lightning. Flaming debris fell on the carpets. Layers of ash descended, alighted, swirled in the wind, and drifted. Molten polyester hardened on the cellar stairs. Almost simultaneously there occurred a major accident in the ice-cream factory next door. As yet no people had arrived. Dead of night. Distant city. And before long the west wall of the House of Carpets fell in under the pressure and weight of a broad, braided ooze of six admixing flavors, which slowly entered Nasser Aftab's show-room and folded and double-folded and covered what was left of his carpets, moving them, as well, some distance across the room. Snow began to fall. It turned to sleet, and soon to freezing rain. In heavy winds under clearing skies, the temperature fell to six below zero. Celsius. Representatives of two warring insurance companies showed up just in front of the fire engines. The insurance companies needed to know precisely what had happened, and in what order, and to what extent it was Aftab's fault. If not a hundred per cent, then to what extent was it the ice-cream factory's fault? And how much fault must be —regrettably—assigned to God? The problem was obviously too tough for the Chicken Valley Police Department, or, for that matter, for any ordinary de-tective. It was a problem, naturally, for a field geolo-gist. One shuffled in eventually. Scratched-up boots. A puzzled look. He picked up bits of wall and ceil-ing, looked under the carpets, tasted the ice cream. He felt the risers of the cellar stairs. Looking up, he told Hartford everything it wanted to know. For him this was so simple it was a five-minute job.

From the high ridges right down to the level of the road, there was snow all over the Ruby Moun-tains. "Ugh," said Deffeyes—his comment on the snow.

"Spoken like a skier," I said.

He said, "I'm a retired skier."

He skied for the School of Mines. In other Rocky Mountain colleges and universities at the time, the best skiers in the United States were duly enrolled and trying to look scholarly and masquerading as amateurs to polish their credentials for the 1952 Olympic Games. Deffeyes was outclassed even on his own team, but there came a day when a great whiteout sent the superstars sprawling on the mountain. Deffeyes' turn for the slalom came late in the afternoon, and just as he was moving toward the gate the whiteout turned to alpenglow, suddenly bringing into focus the well-compacted snow. He shoved off, and was soon bombing. He was not hurting for weight even then. He went down the mountain like an object dropped from a tower. In the end, his time placed him high among the ranking stars.

Now, in the early evening, crossing Independence Valley, Deffeyes seemed scarcely to notice that the white summits of the Ruby Range—above eleven thousand feet, and the highest mountains in this part of the Great Basin—were themselves being reddened with alpenglow. He was musing aloud, for reasons unapparent to me, about the melting points of tin and lead. He was saying that as a general rule material will flow rather than fracture if it is hotter than half of its melting point measured from absolute zero. At room temperature, you can bend tin and lead. They are solid but they flow. Room temperature is more than halfway between absolute zero and the melting points of tin and lead. At room temperature, you cannot bend glass or cast iron. Room temperature is less than halfway from absolute zero to the melting points of iron and glass. "If you go down into the earth here to a depth that about equals the width of one of these fault blocks, the temperature is halfway between absolute zero and the melting point of the rock. The crust is brittle above that point and plastic below it. Where the brittleness ends is the bottom of the tilting fault block, which rests—floats, if you like—in the hot and plastic, slowly flowing lower crust and upper mantle. I think this is why the ranges are so rhythmic. The spacing between them seems to be governed by their depth—the depth of the cold brittle part of the crust. As you cross these valleys from one range to the next, you can sense how deep the blocks are. If they were a lot deeper than their width—if the temperature gradient were different and the cold brittle zone went down, say, five times the surface width—the blocks would not have mechanical freedom. They could not tilt enough to make these mountains. So I suspect the blocks are shallow—about as deep as they are wide. Earthquake history supports this. Only shallow earthquakes have been recorded in the Basin and Range.

At the western edge of Death Valley, there are great convex mountain faces that are called turtlebacks. To me they are more suggestive of whales. You look at them and you see that they were once plastically deformed. I think the mountains have tilted up enough there to be giving us a peek at the original bottom of a block. Death Valley is below sea level. I would bet that if we could scrape away six thousand feet of gravel from these mile-high basins up here what we would see at the base of these mountains would look like the edge of Death Valley. I haven't published this hypothesis. I think it sounds right. I haven't done any field work in Death Valley. I was just lucky enough to be there in 1961 with the guy who first mapped the geology. I have been lucky all through the years to work in the Basin and Range. The Basin and Range impresses me in terms of geology as does no other place in North America. It's not at all easy, anywhere in the province, to say just what happened and when. Range after range—it is mysterious to me. A lot of geology is mysterious to me."

Interstate 80, in its complete traverse of the North American continent, goes through much open space and three tunnels. As it happens, one tunnel passes through young rock, another through middle-aged rock, and the third through rock that is fairly old, at least with respect to the rock now on earth which has not long since been recycled. At Green River, Wyoming, the road goes under a remnant of the bed of a good-sized Cenozoic lake. The tunnel through Yerba Buena Island, in San Francisco Bay, is in sandstones and shales of the Mesozoic. And in Carlin Canyon, in Nevada, the road makes a neat pair of holes in Paleozoic rock. This all but leaves the false impression that an academic geologist chose the sites —and now, as we approached the tunnel at Carlin Canyon, Deffeyes became so evidently excited that one might have thought he had done so himself. "Yewee zink bogawa!" he said as the pickup rounded a curve and the tunnel appeared in view. I glanced at him, and then followed his gaze to the slope above the tunnel, and failed to see there in the junipers and the rubble what it was that could cause this professor to break out in such language. He did not slow up. He had been here before. He drove through the westbound tube, came out into daylight, and, pointing to the right, said, "Shazam!" He stopped on the shoulder, and we admired the scene. The Humboldt River, blue and full, was flowing toward us, with panes of white ice at its edges, sage and green meadow beside it, and dry russet uplands rising behind. I said I thought that was lovely. He said yes, it was lovely indeed, it was one of the loveliest angular unconformities I was ever likely to see.

The river turned in our direction after bending by a wall of its canyon, and the wall had eroded so unevenly that a prominent remnant now stood on its own as a steep six-hundred-foot hill. It made a mammary silhouette against the sky. My mind worked its way through that image, but still I was not seeing what Deffeyes was seeing. Finally, I took it in. More junipers and rubble and minor creases of erosion had helped withhold the story from my eye. The hill, structurally, consisted of two distinct rock formations, awry to each other, awry to the gyroscope of

the earth—just stuck together there like two artistic
impulses in a pointedly haphazard collage. Both
formations were of stratified rock, sedimentary rock,
put down originally in and beside the sea, where
they had lain, initially, flat. But now the strata of the
upper part of the hill were dipping more than sixty
degrees, and the strata of the lower part of the hill
were standing almost straight up on end. It was as if,
through an error in demolition, one urban building
had collapsed upon another. In order to account for
that hillside, Deffeyes was saying, you had to build a
mountain range, destroy it, and then build a second
set of mountains in the same place, and then for the
most part destroy them. You would first have had the
rock of the lower strata lying flat—a conglomerate
with small bright pebbles like effervescent bubbles
in a matrix red as wine. Then the forces that had
compressed the region and produced mountains
would have tilted the red conglomerate, not to the
vertical, where it stood now, but to something like
forty-five degrees. That mountain range wore away
—from peaks to hills to nubbins and on down to
nothing much but a horizontal line, the bevelled surface of slanting strata, eventually covered by a sea.
In the water, the new sediment of the upper formation would have accumulated gradually upon that
surface, and, later, the forces building a fresh mountain range would have shoved, lifted, and rotated the

whole package to something close to its present position, with its lower strata nearly vertical and its upper strata aslant. Here in Carlin Canyon, basin-and-range faulting, when it eventually came along, had not much affected the local structure, further tilting the package only two or three degrees.

Clearly, if you were going to change a scene, and change it again and again, you would need adequate time. To make the rock of that lower formation and then tilt it up and wear it down and deposit sediment on it to form the rock above would require an immense quantity of time, an amount that was expressed in the clean, sharp line that divided the formations—the angular unconformity itself. You could place a finger on that line and touch forty million years. The lower formation, called Tonka, formed in middle Mississippian time. The upper formation, called Strathearn, was deposited forty million years afterward, in late Pennsylvanian time. Cambrian, Ordovician, Silurian, Devonian, Mississippian, Pennsylvanian, Permian, Triassic, Jurassic, Cretaceous, Paleocene, Eocene, Oligocene, Miocene, Pliocene, Pleistocene . . . In the long roll call of the geologic systems and series, those formations—those discrete depositional events, those forty million years —were next-door neighbors on the scale of time. The rock of the lower half of that hill dated to three hundred and thirty million years ago, in the Mississippian, and the rock above the unconformity dated to two hundred and ninety million years ago, in the Pennsylvanian. If you were to lift your arms and spread them wide and hold them straight out to either side and think of the distance from fingertips to fingertips as representing the earth's entire history, then you would have all the principal events in that hillside in the middle of the palm of one hand.

It was an angular unconformity in Scotland— exposed in a riverbank at Jedburgh, near the border, exposed as well in a wave-scoured headland where the Lammermuir Hills intersect the North Sea—that helped to bring the history of the earth, as people had understood it, out of theological metaphor and into the perspectives of actual time. This happened toward the end of the eighteenth century, signalling a revolution that would be quieter, slower, and of another order than the ones that were contemporary in America and France. According to conventional wisdom at the time, the earth was between five thousand and six thousand years old. An Irish archbishop (James Ussher), counting generations in his favorite book, figured this out in the century before. Ussher actually dated the earth, saying that it was created in 4004 B.C. The Irish, as any Oxbridge don would know, are imprecise, and shortly after the publication of Ussher's *Annales Veteris et Novi Testamenti* the Vice-Chancellor of Cambridge University bestirred himself to refine the calculations. He confirmed the year. The Holy Trinity had indeed created the earth in 4004 B.C.—and they had done so, reported the Vice-Chancellor, on October 26th, at 9 A.M. His name was Lightfoot. Geologists today will give parties on the twenty-sixth of October. Some of these parties begin on the twenty-fifth and end at nine in the morning.

It was also conventional wisdom toward the end of the eighteenth century that sedimentary rock had been laid down in Noah's Flood. Marine fossils in mountains were creatures that had got there during the Flood. To be sure, not everyone had always believed this. Leonardo, for example, had noticed fossil clams in the Apennines and, taking into account the distance to the Adriatic Sea, had said, in effect, that it must have been a talented clam that could travel a hundred miles in forty days. Herodotus had seen the Nile Delta—and he had seen in its accumulation unguessable millennia. G. L. L. de Buffon, in 1749 (the year of *Tom Jones*), began publishing his forty-four-volume *Histoire Naturelle*, in which he said that the earth had emerged hot from the sun seventy-five thousand years before. There had been, in short, assorted versions of the Big Picture. But the scientific hypothesis that overwhelmingly prevailed at the time of Bunker Hill was neptunism—the aqueous origins of the visible world. Neptunism had become a systematized physiognomy of the earth, carried forward to the *n*th degree by a German academic mineralogist who published very little but whose teaching was so renowned that his interpretation of the earth was taught as received fact at Oxford and Cambridge, Turin and Leyden, Harvard, Princeton, and Yale. His name was Abraham Gottlob Werner. He taught at Freiberg Mining Academy. He had never been outside Saxony. Extrapolation was his means of world travel. He believed in "universal formations." The rock of Saxony was, beyond a doubt, by extension the rock of Peru. He believed that rock of every kind—all of what is now classified as igneous, sedimentary, and metamorphic—had precipitated out of solution in a globe-engulfing sea. Granite and serpentine, schist and gneiss had precipitated first and were thus "primitive" rocks, the cores and summits of mountains. "Transitional" rocks (slate, for example) had been deposited underwater on high mountain slopes in tilting beds. As the great sea fell and the mountains dried in the sun, "secondary" rocks (sandstone, coal, basalt, and more) were deposited flat in waters above the piedmont. And while the sea kept withdrawing, "alluvial" rock—the "tertiary," as it was sometimes called—was established on what now are coastal plains. That was the earth's surface as it was formed and had remained. There was no hint of where the water went. Werner was gifted with such rhetorical grace that he could successfully omit such details. He could gesture toward the Saxon hills—toward great pyramids of basalt that held castles in the air—and say, without immediate fear of contradiction, "I hold that no basalt is volcanic." He could dismiss volcanism itself as the surface effect of spontaneous combustion of coal. His ideas may now seem risible in direct proportion to their amazing circulation, but that is characteristic more often than not of the lurching progress of science. Those who laugh loudest laugh next. And some contemporary geologists discern in Werner the lineal antecedence of what has come to be known as black-box geology—people in white coats spending summer days in basements watching million-dollar consoles that flash like northern lights—for Werner's "first sketch of a classification of rocks shows by its meagreness how slender at that time was his practical acquaintance with rocks in the field." The words are Sir Archibald Geikie's, and they appeared in 1905 in a book called *The Founders of Geology.* Geikie, director general of the Geological Survey of Great Britain and Ireland, was an accomplished geologist who seems to have dipped in ink the sharp end of his hammer. In summary, he said of Werner, "Through the loyal devotion of his pupils, he was elevated even in his lifetime into the position of a kind of scientific pope, whose decisions were final on any subject regarding which he chose to pronounce them. . . . Tracing in the arrangement of the rocks of the earth's crust the history of an original oceanic envelope, finding in the masses of granite, gneiss, and mica-schist the earliest precipitations from that ocean, and recognising the successive alterations in the constitution of the water as witnessed by the series of geological formations, Werner launched upon the world a bold conception which might well fascinate many a listener to whom the laws of chemistry and physics, even as then understood, were but little known." Moreover, Werner's earth was compatible with Genesis and was thus not unpleasing to the Pope himself. When Werner's pupils, as they spread through the world, encountered reasoning that ran contrary to Werner's, pictures that failed to resemble his picture, they described all these heresies as "visionary fabrics" —including James Hutton's *Theory of the Earth; or, an Investigation of the Laws Observable in the Composition, Dissolution, and Restoration of Land Upon the Globe,* which was first presented before the Royal Society of Edinburgh at its March and April meetings in 1785.

Hutton was a medical doctor who gave up medicine when he was twenty-four and became a farmer who at the age of forty-two retired from the farm. Wherever he had been, he had found himself drawn to riverbeds and cutbanks, ditches and borrow pits, coastal outcrops and upland cliffs; and if he saw black shining cherts in the white chalks of Norfolk, fossil clams in the Cheviot Hills, he wondered why they were there. He had become preoccupied with the operations of the earth, and he was beginning to discern a gradual and repetitive process measured out in dynamic cycles. Instead of attempting to imagine how the earth may have appeared at its vague and unobservable beginning, Hutton thought about the earth as it was; and what he did permit his imagination to do was to work its way from the present moment backward and forward through time. By studying rock as it existed, he thought he could see what it had once been and what it might become. He moved to Edinburgh, with its geologically dramatic setting, and lived below Arthur's Seat and the Salisbury Crags, remnants of what had once been molten rock. It was impossible to accept those battlement hills precipitating in a sea. Hutton had a small fortune, and did not have to distract himself for food. He increased his comfort when he invested in a company that made sal ammoniac from collected soot of the city. He performed experiments— in chemistry, mainly. He extracted table salt from a zeolite. But for the most part—over something like fifteen years—he concentrated his daily study on the building of his theory.

Growing barley on his farm in Berwickshire, he had perceived slow destruction watching streams carry soil to the sea. It occurred to him that if streams were to do that through enough time there would be no land on which to farm. So there must be in the world a source of new soil. It would come from above—that was to say, from high terrain—and be made by rain and frost slowly reducing mountains, which in stages would be ground down from boulders to cobbles to pebbles to sand to silt to mud by a ridge-to-ocean system of dendritic streams. Rivers would carry their burden to the sea, but along the way they would set it down, as fertile plains. The Amazon had brought off the Andes half a continent of plains. Rivers, especially in flood, again and again would pick up the load, to give it up ultimately in depths of still water. There, in layers, the mud, silt, sand, and pebbles would pile up until they reached a depth where heat and pressure could cause them to become consolidated, fused, indurated, lithified— rock. The story could hardly end there. If it did, then the surface of the earth would have long since worn smooth and be some sort of global swamp. "Old continents are wearing away," he decided, "and new continents forming in the bottom of the sea." There were fossil marine creatures in high places. They had not got up there in a flood. Something had lifted the rock out of the sea and folded it up as mountains.

One had only to ponder volcanoes and hot springs to sense that there was a great deal of heat within the earth—much exceeding what could ever be produced by an odd seam of spontaneously burning coal —and that not only could high heat soften up rock and change it into other forms of rock, it could apparently move whole regions of the crustal package and bend them and break them and elevate them far above the sea.

Granite also seemed to Hutton to be a product of great heat and in no sense a precipitate that somehow grew in water. Granite was not, in a sequential sense, primitive rock. It appeared to him to have come bursting upward in a hot fluid state to lift the country above it and to squirt itself thick and thin into preexisting formations. No one had so much as imagined this before. Basalt was no precipitate, either. In Hutton's description, it had once been molten, exhibiting "the liquefying power and expansive force of subterranean fire." Hutton's insight was phenomenal but not infallible. He saw marble as having once been lava, when in fact it is limestone cooked under pressure in place.

Item by item, as the picture coalesced, Hutton did not keep it entirely to himself. He routinely spent his evenings in conversation with friends, among them Joseph Black, the chemist, whose responses may have served as a sort of fixed foot to the wide-swing-ing arcs of Hutton's speculations—about the probable effect on certain materials of varying ratios of temperature and pressure, about the story of the forming of rock. Hutton was an impulsive, highly creative thinker. Black was deliberate and critical. Black had a judgmental look, a lean and sombre look. Hutton had dark eyes that flashed with humor under a far-gone hairline and an oolitic forehead full of stored information. Black is regarded as the discoverer of carbon dioxide. He is one of the great figures in the history of chemistry. Hutton and Black were among the founders of an institution called the Oyster Club, where they whiled away an evening a week with their preferred companions—Adam Smith, David Hume, John Playfair, John Clerk, Robert Adam, Adam Ferguson, and, when they were in town, visitors from near and far such as James Watt and Benjamin Franklin. Franklin called these people "a set of as truly great men . . . as have ever appeared in any Age or Country." The period has since been described as the Scottish Enlightenment, but for the moment it was only described as the Oyster Club. Hutton, who drank nothing, was a veritable cup running over with enthusiasm for the achievements of his friends. When Watt came to town to report distinct progress with his steam engine, Hutton reacted with so much pleasure that one might have thought he was building the thing himself. While the others busied themselves with their economics, their architecture, art, mathematics, and physics, their naval tactics and ranging philosophies, Hutton shared with them the developing fragments of his picture of the earth, which, in years to come, would gradually remove the human world from a specious position in time in much the way that Copernicus had removed us from a specious position in the universe.

A century after Hutton, a historian would note that "the direct antagonism between science and theology which appeared in Catholicism at the time of the discoveries of Copernicus and Galileo was not seriously felt in Protestantism till geologists began to impugn the Mosaic account of the creation." The date of the effective beginning of the antagonism was the seventh of March, 1785, when Hutton's theory was addressed to the Royal Society in a reading that in all likelihood began with these words: "The purpose of this Dissertation is to form some estimate with regard to the time the globe of this Earth has existed." The presentation was more or less off the cuff, and ten years would pass before the theory would appear (at great length) in book form. Meanwhile, the Society required that Hutton get together a synopsis of what was read on March 7th and finished on April 4, 1785. The present quotations are from that abstract.

We find reason to conclude, *1st*, That the land on which we rest is not simple and original, but that it is a composition, and had been formed by the operation of second causes. *2dly*, That before the present land was made there had subsisted a world composed of sea and land, in which were tides and currents, with such operations at the bottom of the sea as now take place. And, *Lastly*, That while the present land was forming at the bottom of the ocean, the former land maintained plants and animals . . . in a similar manner as it is at present. Hence we are led to conclude that the greater part of our land, if not the whole, had been produced by operations natural to this globe; but that in order to make this land a permanent body resisting the operations of the waters two things had been required; *1st*, The consolidation of masses formed by collections of loose or incoherent materials; *2dly*, The elevation of those consolidated masses from the bottom of the sea, the place where they were collected, to the stations in which they now remain above the level of the ocean. . . .

Having found strata consolidated with every species of substance, it is concluded that strata in general have not been consolidated by means of aqueous solution. . . .

It is supposed that the same power of extreme heat by which every different mineral substance had been brought into a melted state might be capable of producing an expansive force sufficient for elevating the land from the bottom of the ocean to the place it now occupies above the surface of the sea. . . .

A theory is thus formed with regard to a mineral system. In this system, hard and solid bodies are to be formed from soft bodies, from loose or incoherent materials, collected together at the bottom of the sea; and the bottom of the ocean is to be made to change its place . . . to be formed into land. . . .

Having thus ascertained a regular system in which the present land of the globe had been first formed at the bottom of the ocean and then raised above the surface of the sea, a question naturally occurs with regard to time; what had been the space of time necessary for accomplishing this great work? . . .

We shall be warranted in drawing the following conclusions; *1st,* That it had required an indefinite space of time to have produced the land which now appears; *2dly,* That an equal space had been employed upon the construction of that former land from whence the materials of the present came; *Lastly,* That there is presently laying at the bottom of the ocean the foundation of future land. . . .

As things appear from the perspective of the twentieth century, James Hutton in those readings became the founder of modern geology. As things appeared to Hutton at the time, he had constructed a theory that to him made eminent sense, he had put himself on the line by agreeing to confide it to the world at large, he had provoked not a few hornets

-like the experimental physicists
go off to check on Einstein by
:dges of solar eclipses—he had
onal travelling to see if he was
:xpress all this in a chapter head-
tely wrote his book, he needed to
nfirmed from Observations made
cidate the Subject." He went to
it to Banffshire. He went to Salt-
Rumbling Bridge. He went to the
Isle of Man, Inchkeith Island in the
His friend John Clerk sometimes
went ... nd made line drawings and water-
colors of scenes that arrested Hutton's attention. In
1968, a John Clerk with a name too old for Roman
numerals found a leather portfolio at his Midlothian
estate containing seventy of those drawings, among
them some cross-sections of mountains with granite
cores. Since it was Hutton's idea that granite was not
a "primary" rock but something that had come up
into Scotland from below, molten, to intrude itself
into the existing schist, there ought to be pieces of
schist embedded here and there in the granite. There
were. "We may now conclude," Hutton wrote later,
"that without seeing granite actually in a fluid state
we have every demonstration possible of this fact;
that is to say, of granite having been forced to flow in
a state of fusion among the strata broken by a sub-

terraneous force, and distorted in every manner and degree."

What called most for demonstration was Hutton's essentially novel and all but incomprehensible sense of time. In 4004 + 1785 years, you would scarcely find the time to make a Ben Nevis, let alone a Gibraltar or the domes of Wales. Hutton had seen Hadrian's Wall running across moor and fen after sixteen hundred winters in Northumberland. Not a great deal had happened to it. The geologic process was evidently slow. To accommodate his theory, all that was required was time, adequate time, time in quantities no mind had yet conceived; and what Hutton needed now was a statement in rock, a graphic example, a breath-stopping view of deep time. There was a formation of "schistus" running through southern Scotland in general propinquity to another formation called Old Red Sandstone. The schistus had obviously been pushed around, and the sandstone was essentially flat. If one could see, somewhere, the two formations touching each other with strata awry, one could not help but see that below the disassembling world lie the ruins of a disassembled world below which lie the ruins of still another world. Having figured out inductively what would one day be called an angular unconformity, Hutton went out to look for one. In a damp country covered with heather, with gorse and bracken, with larches

and pines, textbook examples of exposed rock were
extremely hard to find. As Hutton would write later,
in the prototypical lament of the field geologist, "To
a naturalist nothing is indifferent; the humble moss
that creeps upon the stone is equally interesting as
the lofty pine which so beautifully adorns the valley
or the mountain: but to a naturalist who is reading in
the face of rocks the annals of a former world, the
mossy covering which obstructs his view, and ren-
ders undistinguishable the different species of stone,
is no less than a serious subject of regret." Hutton's
perseverance, though, was more than equal to the
irksome vegetation. Near Jedburgh, in the border
country, he found his first very good example of an
angular unconformity. He was roaming about the
region on a visit to a friend when he came upon a
stream cutbank where high water had laid bare the
flat-lying sandstone and, below it, beds of schistus
that were standing straight on end. His friend John
Clerk later went out and sketched for Hutton this
clear conjunction of three worlds—the oldest at the
bottom, its remains tilted upward, the intermediate
one a flat collection of indurated sand, and the
youngest a landscape full of fences and trees with a
phaeton-and-two on a road above the rivercut, driver
whipping the steeds, rushing through a moment in
the there and then. "I was soon satisfied with regard
to this phenomenon," Hutton wrote later, "and re-

joiced at my good fortune in stumbling upon an object so interesting to the natural history of the earth, and which I had been long looking for in vain."

What was of interest to the natural history of the earth was that, for all the time they represented, these two unconforming formations, these two levels of history, were neighboring steps on a ladder of uncountable rungs. Alive in a world that thought of itself as six thousand years old, a society which had placed in that number the outer limits of its grasp of time, Hutton had no way of knowing that there were seventy million years just in the line that separated the two kinds of rock, and many millions more in the story of each formation—but he sensed something like it, sensed the awesome truth, and as he stood there staring at the riverbank he was seeing it for all mankind.

To confirm what he had observed and to involve further witnesses, he got into a boat the following spring and went along the coast of Berwickshire with John Playfair and young James Hall, of Dunglass. Hutton had surmised from the regional geology that they would come to a place among the terminal cliffs of the Lammermuir Hills where the same formations would touch. They touched, as it turned out, in a headland called Siccar Point, where the strata of the lower formation had been upturned to become vertical columns, on which rested the Old Red Sandstone, like the top of a weather-beaten table. Hutton, when he eventually described the scene, was both gratified and succinct—"a beautiful picture . . . washed bare by the sea." Playfair was lyrical:

On us who saw these phenomena for the first time, the impression made will not easily be forgotten. The palpable evidence presented to us, of one of the most extraordinary and important facts in the natural history of the earth, gave a reality and substance to those theoretical speculations, which, however probable, had never till now been directly authenticated by the testimony of the senses. We often said to ourselves, What clearer evidence could we have had of the different formation of these rocks, and of the long interval which separated their formation, had we actually seen them emerging from the bosom of the deep? We felt ourselves necessarily carried back to the time when the schistus on which we stood was yet at the bottom of the sea, and when the sandstone before us was only beginning to be deposited, in the shape of sand or mud, from the waters of a superincumbent ocean. An epocha still more remote presented itself, when even the most ancient of these rocks, instead of standing upright in vertical beds, lay in horizontal planes at the bottom of the sea, and was not yet disturbed by that immeasurable force which has burst asunder the solid pavement of the globe. Revolutions still more remote appeared in the distance of this extraordinary perspective. The mind seemed to grow giddy by looking so far into the abyss of time.

Hutton had told the Royal Society that it was his purpose to "form some estimate with regard to the time the globe of this Earth has existed." But after Jedburgh and Siccar Point what estimate could there be? "The world which we inhabit is composed of the materials not of the earth which was the immediate predecessor of the present but of the earth which . . . had preceded the land that was above the surface of the sea while our present land was yet beneath the water of the ocean," he wrote. "Here are three distinct successive periods of existence, and each of these is, in our measurement of time, a thing of indefinite duration. . . . The result, therefore, of this physical inquiry is, that we find no vestige of a beginning, no prospect of an end."

The Old Red Sandstone was put down by rivers flowing southward to a sea where marine strata were accumulating in the region that is now called Devon. The size, speed, and direction of the rivers—their islands, pitches, and bends—are not just inferable but can almost be seen, in structures in the Old Red Sandstone: gravel bars, point bars, ripples of the riverbeds, migrating channels, "waves" that formed of sand. The sea into which those rivers spilled ran all the way to Russia, but it was in the rock of Devonshire that geologists in the eighteen-thirties found cup corals—fossilized skeletons, cornucopian in shape—that were not of an age with corals they had found before. They had found related corals that were obviously less developed than these, and they had found corals that were more so.

The less developed corals had been in rock that lay under the Old Red Sandstone. The more developed corals had been in rock above the Old Red Sandstone. Therefore, it was inferred (correctly) that the Old Red Sandstone of North Britain and the marine limestone of Devon were of the same age, and that henceforth any rock of that age anywhere in the world—in downtown Iowa City; on Pequop Summit, in Nevada; in Stroudsburg, Pennsylvania; in Sandusky, Ohio—would be called Devonian. It was a name given, although they did not know it then, to fifty million years. They still had no means of measuring the time involved. They also had no way of knowing that those fifty million years had ended a third of a billion years ago. All they had was their new and expanding insight that they were dealing with time in quantities beyond comprehension. Devonian—395 to 345 million years before the present.

Geologists did not have to look long at the coal seams of Europe—the coals of the Ruhr, the coals of the Tyne—to decide that the coals were of an age, which they labelled Carboniferous. The coal and related strata lay on top of the Old Red Sandstone. So, in the succession of time, the Carboniferous period (eventually subdivided into Mississippian and Pennsylvanian in the United States) would follow the Devonian, coupling on, as the science would even-

tually determine, another sixty-five million years—
345 to 280 million years before the present.

In this manner—with their fossil assemblages
and faunal successions, their hammers decoding rock
—geologists in the first eighty years of the nineteenth century constructed their scale of time. It was
based on organic evolution, and, crossing the century, it both anticipated and confirmed Darwin.
When the Devonian was defined in the light of the
changes in corals, Darwin was obscure and not long
off the Beagle, with twenty years to go before *The
Origin of Species*. Meanwhile, the geologists were
out correlating strata and reading there a record less
of rock than of life. The rock had been recycled, and
sandstones of one era could be indistinguishable
from the sandstones of another, but organic evolution had not occurred in cycles, so it was through the
antiquity of fossils that geologists worked out the
comparative ages of the rock in which the fossils
were preserved. Some creatures were more useful
than others. Oysters and horseshoe crabs, for example, were of marginal assistance. Oysters had appeared in the Triassic, horseshoe crabs in the Cambrian. Both had evolved minimally and had
obviously avoided extinction. Some creatures, on the
other hand, had appeared suddenly, had evolved
quickly, had become both abundant and geographi-
cally widespread, and then had died out, or died

down, abruptly. Geologists canonized them as "index fossils" and studied them in groups. Experience proved that the surest method of working out relative ages of rock was not through individual creatures but through the relating of successive strata to whole collections of creatures whose fossils were contained therein—a painstaking comparison of arrivals and extinctions that helped to characterize the divisions of the time scale and define its boundaries with precision.

Imagine an E. L. Doctorow novel in which Alfred Tennyson, William Tweed, Abner Doubleday, Jim Bridger, and Martha Jane Canary sit down to a dinner cooked by Rutherford B. Hayes. Geologists would call that a fossil assemblage. And, without further assistance from Doctorow, a geologist could quickly decide—as could anyone else—that the dinner must have occurred in the middle eighteen-seventies, because Canary was eighteen when the decade began, Tweed became extinct in 1878, and the biographies of the others do not argue with these limits. In progressive refinements, geologists with their fossil assemblages established their systems and series and stages of rock, their eras and periods and epochs of time. But, unlike Doctorow, who deals with a mere half-dozen people around a dinner table, the geologists would assemble from one set of strata hundreds and even thousands of species from all over the food chain, and by lining up their genetic histories side by side establish with near-certainty points in comparative time.

Some of these time lines were bolder than others, and none more so than the one that underlined the first appearance of fossils in abundance in the world. It marked a great and sudden explosion of life, all the major phyla having developed more or less at the same time and now acquiring skeletons and shells and teeth and other hard components that allowed them individually to be reported to the future. Because rock that held these early fossils was first studied on Harlech Dome and adjacent Welsh terrains, geologists named the system Cambrian, after the Roman name for Wales. They then named the Silurian for a Welsh tribe that bitterly defied the Romans. After some years and more comparative study, an argument broke out over the Cambro-Silurian line, a scientific battle royal in which the Cambrian forces tried to move their banner forward through time and the Silurian proponents attempted to push theirs back. The disputed block of time became a sort of demilitarized zone. Friendships came unstuck. The standoff lasted for decades, until some genius in scientific diplomacy suggested that the disputed time had enough characteristics of its own to be given the status of a discrete period, an appropriate name for which—in honor of another tribe of

intractable Welsh belligerents—would be Ordovician.
There was a lot of room for generosity. There was
plenty of time for all. Cambrian—570 to 500. Ordovician—500 to 435. Silurian—435 to 395 million
years before the present.

A British geologist went to Russia and after a
season or two's tapping at the Urals named still another period in time, and system of rock, for the upland oblast of Perm. There were formations in Perm
with a fossil story distinctly their own that were
superimposed—as they happen to be in Pennsylvania, as they happen to be at the rim of the Grand
Canyon—upon the Carboniferous. What was distinct
about the character of the Permian assemblages was
not only the forms to which they had evolved but
also their absence in great numbers from higher,
younger strata. There had evidently been a wave of
death, in which thousands of species had vanished
from the world. No one has explained what happened—at least not to the general satisfaction. A
drastic retreat of shallow seas may have destroyed
innumerable environments. A change of ocean salin-
ity may have ended a lot of life through osmotic
shock. The cause may have been extraterrestrial—
lethal radiation from a supernova dying nearby.
None of these hypotheses has attracted enough con-
currence to be dressed out in full as a theory, but,
whatever the cause, no one argues that at least half

the fish and invertebrates and three-quarters of all amphibians—perhaps as much as ninety-six per cent of all marine faunal species—disappeared from the world in what has come to be known as the Permian Extinction.

It was an extinction of a magnitude that would be approached only once in subsequent history, or— to express that more gravely—only once before the present day. The sharp line of creation at the outset of the Cambrian had an antiphonal parallel in the Permian Extinction, and the whole long stretch between the one and the other was set apart in history as the Paleozoic era. It was a unit—well below the surface but far above the bottom—just hanging there suspended in the formless pelagics of time. No vestige of a beginning. No prospect of an end. The Paleozoic—570 to 230 million years before the present, a thirteenth of the history of the earth. Cambrian, Ordovician, Silurian, Devonian, Mississippian, Pennsylvanian, Permian. When I was seventeen, I used to accordion-pleat those words, mnemonically capturing the vanished worlds of "Cosdmpp," the order of the periods, the sequence of the systems. It was either that or write them in the palm of one hand.

Lyell, Cuvier, Conybeare, Phillips, von Alberti, von Humboldt, Desnoyers, d'Halloy, Sedgwick, Murchison, Lapworth, Smith (William "Strata"

Smith): the geologists who extended Hutton's insight and built this time scale conjoined their names in the history of the science in a way that would not be repeated for more than a hundred years, until a roster of comparable length—Hess, Heezen, McKenzie, Morgan, Wilson, Matthews, Vine, Parker, Sykes, Ewing, Le Pichon, Cox, Menard—would effect the plate-tectonics revolution. The system of rock immediately above the Paleozoic, in which all that Permian life failed to reappear, was typified by three formations in Germany—certain sandstones, limestones, and marly shales—that ran like a striped flag through the Black Forest, the Rhine Valley, and lent the name Triassic to thirty-five million years. In the Triassic, the earliest subdivision of the Mesozoic era, two families of reptiles that had survived the Permian Extinction began to show patterns of unprecedented growth. This would continue for a hundred and fifty million years—through the Jurassic and out to the end of Cretaceous time, when the "fearfully great lizards," on the point of disappearance, would reach their greatest size, not to be surpassed until epochs that followed the Eocene development of whales. European geologists studying the massive limestones of the Jura—the gentle mountains of the western cantons of Switzerland and of Franche-Comté—related the copious displays of ancient life there to comparable assemblages elsewhere

in the world, and called them all Jurassic. The first
bird appeared in the Jurassic. It had claws on its
wings and teeth in its bill and a reptile's long tail
sprouting feathers. Its complete performance envelope as a flier was to climb a tree and jump.

Physicists, chemists, and mathematicians, taking
note of all the nomenclatural inconsistencies—of
time named for mountain ranges, time named for
savage tribes, time named for a country here, a
county there, an oblast in the Urals—have politely,
gently, suggested that, in this one sense only, the
time scale seems archaic, seems, if one may say so,
out of date. Geology might be better served by a
straightforward system of numbers. The reaction of
geologists, by and large, has been to look upon this
suggestion as if it had come over a bridge that exists
between two cultures. A Continental geologist, in
1822, named seventy-six million years for the white
cliffs of Dover, for the downs of Kent and Sussex, for
the chalky ground of Cognac and Champagne. Related strata were spread out through Holland,
Sweden, Denmark, Germany, and Poland. He
called it Le Terrain Crétacé. If that name was apt,
his own was irresistible. He was J. J. d'Omalius
d'Halloy. Triassic, Jurassic, Cretaceous. When the
Cretaceous ended, the big marine reptiles had disappeared, the flying reptiles, the dinosaurs, the rudistid
clams, and many species of fish, not to mention the

total elimination or severe reduction of countless smaller species from the sea. Once in a great while, the earth moves through cosmic dust that collects near the arms of the galaxy. It has been suggested that this dust might have deflected enough sunlight to bring on the biotic catastrophe. An ocean gone stagnant would have done the same, killing phytoplankton, which prosper in the currents of mixed-up seas. Break the food chain and creatures die out above the break. Phytoplankton are the base of the food chain. The Arctic Ocean, surrounded by continents that had drifted together, might have become in the Cretaceous the greatest lake in all eternity, and when the North Atlantic opened up enough to let the water flood the southern seas the life in them would have suffered a cold osmotic shock. Drastic fluctuations of sea level—also related, perhaps, to the separation of continents—might have caused changes in air temperature and ocean circulation that were enough to sunder the food chain. At the end of 1979, a small group at the Lawrence Radiation Laboratory, in Berkeley—among them the physicist Luis Alvarez, winner of a Nobel Prize, and his son Walter, who is a geologist—brought forth a piece of science in which they present the catastrophe as the effect of an Apollo Object colliding with the earth. An Apollo Object is an "earth-orbit-crossing" asteroid that is at least a kilometre in diameter and is in the category of asteroids that have pockmarked the surface of Mercury, Mars, and the moon, and the surface of the earth as well, although most of the evidence has been obscured here by erosion. Like the general run of meteorites, an Apollo Object could be expected to contain a percentage of iridium and other platinum-like metals at least a thousand times greater than the concentration of the same metals in the crust of the earth. In widely separated parts of the world—Italy, Denmark, New Zealand—the Berkeley researchers have found a thin depositional band, often just a centimetre thick, that contains unearthly concentrations of iridium. Below that sharp line are abundant Cretaceous fossils, and above it they are gone. It marks precisely the end of Cretaceous time. The Berkeley calculations suggest an asteroid about six miles in diameter hitting the earth with a punch of a hundred million megatons, making a crater a hundred miles wide. Such an occurrence—which could repeat itself tomorrow afternoon, there being something like seven hundred big asteroids out there in threatening orbits—would have sent up a mushroom cloud containing some thirty thousand cubic kilometres of pulverized asteroid and terrestrial crust, part of which would have gone into the stratosphere and spread quickly over the earth, keeping sunlight off the lands and seas and suppressing photosynthesis. On August 26 and 27,

1883, when the island Krakatoa, in the Sunda Strait, exploded with great violence, it sent less than twenty cubic kilometres of material into the air, but within a few days dust had spread above the whole earth, turning daylight into dusk. It made exceptionally brilliant sunsets for two and a half years. Edmund Halley, who died when James Hutton was fifteen, once wrote a paper suggesting that the way God started Noah's Flood was by directing a big comet into collision with the earth. The Cretaceous Extinction, whatever its cause, was one of the two most awesome annihilations of life in the history of the world. With the Permian Extinction before it, it framed the Mesozoic, an era of burgeoning creation within deadly brackets of time.

For establishing our bearings through time, we obviously owe an incalculable debt to vanished and endangered species, and if the condor, the kit fox, the human being, the black-footed ferret, and the three-toed sloth are at the head of the line to go next, there is less cause for dismay than for placid acceptance of the march of prodigious tradition. The opossum may be Cretaceous, certain clams Devonian, and oysters Triassic, but for each and every oyster in the sea, it seems, there is a species gone forever. Be a possum is the message, and you may outlive God. The Cenozoic era—coming just after the Cretaceous Extinction, and extending as it does to the

latest tick of time—was subdivided in the eighteen-
thirties according to percentages of molluscan species that have survived into the present. From the
Eocene, for example, which ended some thirty-eight
million years ago, roughly three and a half per cent
have survived. Eocene means "dawn of the recent."
The first horse appeared in the Eocene. Looking
something like a toy collie, it stood three hands high.
From the Miocene ("moderately recent"), some fifteen per cent of molluscan species survive; from the
Pliocene ("more recent"), the number approaches
half. As creatures go, mollusks have been particularly hardy. Many species of mammals fell in the
Pliocene as prairie grassland turned to tundra and
ice advanced from the north. From the Pleistocene
("most recent"), more than ninety per cent of molluscan species live on. The Pleistocene has also been
traditionally defined by four great glacial pulsations,
spread across a million years—the Nebraskan ice
sheet, the Kansan ice sheet, the Illinoian and Wisconsinan ice sheets. It now appears that these were
the last of many glacial pulsations that have occurred in relatively recent epochs, beginning probably in the Miocene and reaching a climax in the ice
sheets of Pleistocene time. The names of the Cenozoic epochs were proposed by Charles Lyell, whose
Principles of Geology was the standard text through
much of the nineteenth century. To settle problems

here and there, the Oligocene ("but a little recent")
was inserted in the list, and the Paleocene ("old recent") was sliced off the beginning. Paleocene,
Eocene, Oligocene, Miocene, Pliocene, Pleistocene—
sixty-five million to ten thousand years before the
present. Divisions grew shorter in the Cenozoic—the
epochs range from eighteen million years to less than
two million—because so much remains on earth of
Cenozoic worlds.

Ignoring its geology, I guess I don't know a
paragraph in literature that I prefer to the one Joseph Conrad begins by saying, "Going up that river
was like travelling back to the earliest beginnings
of the world, when vegetation rioted on the earth
and the big trees were kings." He says, moments
later, "This stillness of life did not in the least resemble a peace. It was the stillness of an implacable
force brooding over an inscrutable intention. It
looked at you with a vengeful aspect. I got used to it
afterwards; I did not see it anymore; I had no time. I
had to keep guessing at the channel; I had to discern,
mostly by inspiration, the signs of hidden banks; I
watched for sunken stones." Metaphorically, he travelled back to the Carboniferous, when the vegetal
riot occurred, but scarcely was that the beginning of
the world. The first plants to appear on land, ever,
appeared in the Silurian. Through the Ordovician
and the Cambrian, there had been no terrestrial veg-

etation at all. And in the deep shadow below the Cambrian were seven years for every one in all subsequent time. There were four billion years back there—since the earliest beginnings of the world. There were scant to nonexistent fossils. There were the cores of the cratons, the rock of the continental shields, the rock of the surface of the moon. There were the reefs of the Witwatersrand. There was the rock that would become the Adirondack Mountains, the Wind River summits, the Seward Peninsula, Manhattan Island. But so little is known of this seven-eighths of all history that in a typical two-pound geological textbook there are fourteen pages on Precambrian time. The Precambrian has attracted geologists of exceptional imagination, who see families of mountains in folded schists. Uranium-lead and potassium-argon radiometric dating have helped them to sort out their Kenoran, Hudsonian, Elsonian Orogenies, their Aphebian, Hadrynian, Paleohelikian time. Isolating the first two billion years of the life of the earth, they called it the Archean Eon. In the Middle Archean, photosynthesis began. Much later in the Precambrian, somewhere in Helikian or Hadrynian time, aerobic life appeared. There is no younger rock in the United States than the travertine that is forming in Thermopolis, Wyoming. A 2.7-billion-year-old outcrop of the core of the continent is at the head of Wind River Canyon, twenty miles away. Precambrian—4,600 to 570 million years before the present.

At the other end of the scale is the Holocene, the past ten thousand years, also called the Recent—Cro-Magnon brooding beside the melting ice. (The Primitive and Secondary eras of eighteenth-century geology are long since gone from the vocabulary, but oddly enough the Tertiary remains. The term, which is in general use, embraces nearly all of the Cenozoic, from the Cretaceous Extinction to the end of the Pliocene, while the relatively short time that follows—the Pleistocene plus the Holocene—has come to be called the Quaternary. The moraines left by ice sheets are Quaternary, as are the uppermost basin fillings in the Basin and Range.) It was at some moment in the Pleistocene that humanity crossed what the geologist-theologian Pierre Teilhard de Chardin called the Threshold of Reflection, when something in people "turned back on itself and so to speak took an infinite leap forward. Outwardly, almost nothing in the organs had changed. But in depth, a great revolution had taken place: consciousness was now leaping and boiling in a space of super-sensory relationships and representations; and simultaneously consciousness was capable of perceiving itself in the concentrated simplicity of its faculties. And all this happened for the first time." Friars of another sort—evangelists of the so-called Environmental Movement—have often made use of the geologic time scale to place in perspective that great "leap forward" and to suggest what our reflective capacities may have meant to Mother Earth. David Brower, for example, the founder of Friends of the Earth and emeritus hero of the Sierra Club, has tirelessly travelled the United States for thirty years delivering what he himself refers to as "the sermon," and sooner or later in every talk he invites his listeners to consider the six days of Genesis as a figure of speech for what has in fact been four and a half billion years. In this adjustment, a day equals something like seven hundred and fifty million years, and thus "all day Monday and until Tuesday noon creation was busy getting the earth going." Life began Tuesday noon, and "the beautiful, organic wholeness of it" developed over the next four days. "At 4 P.M. Saturday, the big reptiles came on. Five hours later, when the redwoods appeared, there were no more big reptiles. At three minutes before midnight, man appeared. At one-fourth of a second before midnight, Christ arrived. At one-fortieth of a second before midnight, the Industrial Revolution began. We are surrounded with people who think that what we have been doing for that one-fortieth of a second can go on indefinitely. They are considered normal, but they are stark raving mad." Brower holds up a photograph of the world—blue, green, and swirling white. "This is the sudden insight from Apollo," he says. "There it is. That's all. We see through the eyes of the astronauts how fragile our life really is." Brower has computed that we are driving through the earth's resources at a rate comparable to a man's driving an automobile a hundred and twenty-eight miles an hour—and he says that we are accelerating.

In like manner, geologists will sometimes use the calendar year as a unit to represent the time scale, and in such terms the Precambrian runs from New Year's Day until well after Halloween. Dinosaurs appear in the middle of December and are gone the day after Christmas. The last ice sheet melts on December 31st at one minute before midnight, and the Roman Empire lasts five seconds. With your arms spread wide again to represent all time on earth, look at one hand with its line of life. The Cambrian begins in the wrist, and the Permian Extinction is at the outer end of the palm. All of the Cenozoic is in a fingerprint, and in a single stroke with a medium-grained nail file you could eradicate human history. Geologists live with the geologic scale. Individually, they may or may not be alarmed by the rate of exploitation of the things they discover, but, like the environmentalists, they use these repetitive analogies to place the human record in perspective—to see the Age of Reflection, the last few thousand years, as a small bright sparkle at the end of time. They often liken humanity's presence on earth to a brief visitation from elsewhere in space, its luminous, explosive characteristics consisting not merely of the burst of population in the twentieth century but of the whole millennial moment of people on earth—a single detonation, resembling nothing so much as a nuclear implosion with its successive neutron generations, whole generations following one another once every hundred-millionth of a second, temperatures building up into the millions of degrees and stripping atoms until bare nuclei are wandering in electron seas, pressures building up to a hundred million atmospheres, the core expanding at five million miles an hour, expanding in a way that is quite different from all else in the universe, unless there are others who also make bombs.

The human consciousness may have begun to leap and boil some sunny day in the Pleistocene, but the race by and large has retained the essence of its animal sense of time. People think in five generations —two ahead, two behind—with heavy concentration on the one in the middle. Possibly that is tragic, and possibly there is no choice. The human mind may not have evolved enough to be able to comprehend deep time. It may only be able to measure it. At least, that is what geologists wonder sometimes, and they have imparted the questions to me. They wonder to what extent they truly sense the passage of millions of years. They wonder to what extent it is possible to absorb a set of facts and move with them, in a sensory manner, beyond the recording intellect and into the abyssal eons. Primordial inhibition may stand in the way. On the geologic time scale, a human lifetime is reduced to a brevity that is too inhibiting to think about. The mind blocks the information. Geologists, dealing always with deep time, find that it seeps into their beings and affects them in various ways. They see the unbelievable swiftness with which one evolving species on the earth has learned to reach into the dirt of some tropical island and fling 747s into the sky. They see the thin band in which are the all but indiscernible stratifications of Cro-Magnon, Moses, Leonardo, and now. Seeing a race unaware of its own instantaneousness in time, they can reel off all the species that have come and gone, with emphasis on those that have specialized themselves to death.

In geologists' own lives, the least effect of time is that they think in two languages, function on two different scales.

"You care less about civilization. Half of me gets upset with civilization. The other half does not get upset. I shrug and think, So let the cockroaches take over."

"Mammalian species last, typically, two million years. We've about used up ours. Every time Leakey finds something older, I say, 'Oh! We're overdue.' We will be handing the dominant-species-on-earth position to some other group. We'll have to be clever not to."

"A sense of geologic time is the most important thing to suggest to the nongeologist: the slow rate of geologic processes, centimetres per year, with huge effects, if continued for enough years."

"A million years is a short time—the shortest worth messing with for most problems. You begin tuning your mind to a time scale that is the planet's time scale. For me, it is almost unconscious now and is a kind of companionship with the earth."

"It didn't take very long for those mountains to come up, to be deroofed, and to be thrust eastward. Then the motion stopped. That happened in maybe ten million years, and to a geologist that's really fast."

"If you free yourself from the conventional reaction to a quantity like a million years, you free yourself a bit from the boundaries of human time. And then in a way you do not live at all, but in another way you live forever."

One is tempted to condense time, somewhat glibly—to say, for example, that the faulting which lifted up the mountains of the Basin and Range began "only" eight million years ago. The late Miocene was "a mere" eight million years ago. That the Rocky Mountains were building seventy million years ago and the Appalachians were folding four hundred million years ago does not impose brevity on eight million years. What is to be avoided is an abridgment of deep time in a manner that tends to veil its already obscure dimensions. The periods are so long—the seventy-six million years of the Cretaceous, the fifty million years of the Devonian—that each has acquired its own internal time scale, intricately constructed and elaborately named. I will not attempt to reproduce this amazing list but only to suggest its profusion. The stages and ages, as they are called—the subdivisions of all of the epochs and eras—read like a roll call in a district council somewhere in Armenia. Berriasian, Valanginian, Hauterivian, Barremian, Bedoulian, Gargasian, Aptian, Albian, Cenomanian, Turonian, Coniacian, Santonian, Campanian, and Maastrichtian, reading upward, are chambers of Cretaceous time. Actually, the Cretaceous has been cut even finer, with about fifty clear time lines now, subdivisions of the subdivisions of its seventy-six million years. The Triassic consists of the Scythian, the Anisian, the Ladinian, the Carnian, the Norian, and the Rhaetian, averaging six million years. What survived the Rhaetian lived on into the Liassic. The Liassic, an epoch, comes just after the Triassic and is the early part of the Jurassic. Kazanian, Couvinean, Kopaninian, Kimmeridgian, Tremadocian, Tournaisian, Tatarian, Tiffanian . . . When geologists choose to ignore these names, as they frequently do, they resort to terms that are undecipherably simple, and will note, typically, that an event which occurred in some flooded summer 330.27 million years ago took place in the "early late-middle Mississippian." To say "middle Mississippian" might do, but with millions of years in the middle Mississippian there is an evident compunction to be more precise. "Late" and "early" always refer to time. "Upper" and "lower" refer to rock. "Upper Devonian" and ·"lower Jurassic" are slices of time expressed in rock.

In the late-middle Mississippian, there was an age called Meramecian, of about six million years, and it was during the Meramecian that the Tonka—the older of the formations in the angular unconformity in Carlin Canyon, Nevada—was accumulating along an island coast. The wine-red sandstone and its pebbles may have been sand and pebbles of the beach. The island was of considerable size, apparently, and stood off North America in much the way that Taiwan now reposes near the coast of China. Where there were swamps, they were full of awkward amphibians, not entirely masking in their appearance the human race they would become. They struggled along on stumpy legs. The strait separating the Meramecian island from the North American mainland was about four hundred miles wide and contained crossopterygian fish, from which the amphibians had evolved. There were shell-crushing sharks, horn corals, meadows of sea lilies, and spiral bryozoans that looked like screws. The strait was warm and equatorial. The equator ran through the present site of San Diego, up through Colorado and Nebraska, and on through the site of Lake Superior. The lake would not be dug for nearly three hundred and thirty million years. If in the Meramecian you were to have followed the present route of Interstate so moving east, you would have raised the coast of North America near the Wyoming border, and landed on a red beach. Gradually, you would have ascended through equatorial fern forests, in red soil, to a high point somewhere near Laramie, to begin there a long general downgrade among low hills to Grand Island, Nebraska, where you would have come to an arm of the sea. The far shore was four hundred miles to the east, where the Mississippi River is now, and beyond it was a low, wet, humid, flat terrain, dense with ferns and fern trees—Illinois, Indiana, Ohio. Halfway across Ohio, you would have come to a second epicratonic sea, its far shore in central Pennsylvania. In New Jersey, you would have begun to ascend mountains and ever higher mountains, their summits girt with ice and capped with snow, not unlike Mt. Kenya, not unlike the present peaks of New Guinea and Ecuador, with their snowfields and glaciers in the equatorial tropics. Reaching the site of the George Washington Bridge, you would have been at considerable altitude, looking at mountains and more mountains before you in future Africa.

If you had turned around and gone back to Nevada a million years later, still in Meramecian time, there would have been few variations to note along the way. The west coast would have moved east, but only a bit, and would still be approximately at the western end of Wyoming. There would have been a significant alteration, however, in the demeanor of the island over the strait. At a little over two inches a year, it would have moved forty miles or so eastward, compressing the floor of the strait and pushing up high mountains, like the present mountains of Timor, which have come up in much the same way to stand ten thousand feet above the Banda Sea. Up from the sea and within those Meramecian Nevada mountains came the wine-red pebbly sandstone of the Tonka Formation.

Forty million years after that, when the Tonka mountains had been worn flat and the Strathearn limestones were forming over their roots, the American scene was very different. It was now the Missourian age of late Pennsylvanian time (about two hundred and ninety million years ago), and the Appalachians were still high but they were no longer alpine. Travelling west, and coming down from the mountains around Du Bois, Pennsylvania, you would have descended into a densely vegetal swamp. This was Pennsylvania in the Pennsylvanian, when vegetation rioted on the earth and the big trees were kings. They were not huge by our standards but they were big trees, some with diamond patterns precisioned in their bark. They had thick boles and were about a hundred feet high. Other trees had bark like the bark of hemlocks and leaves like flat straps. Others had the fluted, swollen bases of cypress. In and out among the trunks flew dragonflies with the wingspans of great horned owls. Amphibians not only were walking around easily but some of them had become reptiles. Through the high meshing crowns of the trees not a whole lot of light filtered down. The understory was all but woven—of rushlike woody plants and seed ferns. There were luxuriant tree ferns as much as fifty feet high. The scene suggests a tropical rain forest but was more akin to the Everglades, the Dismal Swamp, the Atchafalaya basin—a hummocky spongy landscape ending in a ragged coast. All through Pennsylvanian time, ice sheets had been advancing over the southern continents, advancing and retreating, forming and melting, lowering and raising the level of the sea, and as the sea came up and over the land in places like the swamps of Du Bois it buried them, first under beach sand and later—as the seawater deepened—under lime muds. With enough burial, the muds became limestone, the sands became sandstone, the vegetation coal. When the sea fell, erosion wore away some of that, but then the sea would rise again to bury new generations of ferns and trees under successive layers of rock. These cyclothems, as they are called, contain the coals of Pennsylvania, and similar ones the coals of Iowa and Illinois. The shallow sea that reached into western Pennsylvania and eastern Ohio was a hundred miles wide in the Missourian age of Pennsylvanian time, and after crossing the water you would have reached a beach and another coal swamp and then, in light-gray soil, a low lush tropical forest that went on through Indiana to eastern Illinois, where it ended with more coal swamps, another sea. The far shore was where the Mississippi River is now, and beyond that was an equatorial rain forest, which ended in central Iowa with another swamp, another sea. The water here was clear and sparkling, with almost no land-derived sediments settling into it, just accumulating skeletons—clean deep beds of lime. Five hundred miles over the water, you would have raised the rose-colored beaches of eastern Wyoming. Mountains stood out to the south. They were the Ancestral Rockies, and time would bevel them to stumps. Skirting them, in Pennsylvanian Wyoming, you would have traversed what seem to have been Saharan sands, wave after wave of dunal sands, five hundred miles of rose and amber pastel sands, ending at the west coast of North America, in Salt Lake City. As the Pennsylvanian sea level moved up and down here, it left alternating beds of lime and sand, which, two eras later, the nascent Oquirrh Mountains would lift to view. Two hundred miles out to sea was the site of Carlin Canyon, where muds of clean lime were settling. The Strathearn, the younger formation of the two in the Carlin unconformity, is an almost pure limestone.

The two formations, conjoined, were driven upward, according to present theory, in a collision of crustal plates that occurred in the early Triassic. The result was yet another set of new mountains—alpine mountains which erosion brought down before the end of the Jurassic, but not enough to obliterate the story that is told in Carlin Canyon. Still gazing at the Carlin unconformity, Ken Deffeyes said, "Profound as all the time is to build and destroy those mountain ranges, it is just a one-acter in the history of the Basin and Range—small potatoes, weak beer, just a little piece of time, a little piece of the action, lost in all the welter of all the other history." There had been two complete cycles of erosion and deposition and mountain building in this one place in one-fortieth of the time scale. That is what made John Playfair's mind grow giddy when James Hutton took him in 1788 to see the angular unconformity at Siccar Point. It was especially fortunate that Playfair was there, and that Playfair knew Hutton and Hutton's geology equally well, for when Hutton finally wrote his book most readers were trampled by the prose. Hutton was at best a difficult writer. Insights came to him but phrases did not. James Hall, who was twenty-seven when he went with Hutton and Playfair to Siccar Point, would say of Hutton years later, "I must own that on reading Dr. Hutton's first geological publication I was induced to reject his system entirely, and should probably have continued still to do so, with the great majority of the world, but for my habits of intimacy with the author, the vivacity and perspicuity of whose conversation formed a striking contrast to the obscurity of his writings." Hall, incidentally, melted rock in crucibles and saw how crystals formed as it cooled. He is regarded as the founder of experimental geology. John Playfair likewise assessed Hutton's literary style as containing "a degree of obscurity astonishing to those who knew him, and who heard him every day converse with no less clearness and precision than animation and force." One can imagine what Playfair thought when he read something like this in Hutton's two-volume *Theory of the Earth*:

If, in examining our land, we shall find a mass of matter which had been evidently formed originally in the ordinary manner of stratification, but which is now extremely distorted in its structure and displaced in its position,— which is also extremely consolidated in its mass and variously changed in its composition,—which therefore has the marks of its original or marine composition extremely obliterated, and many subsequent veins of melted mineral matter interjected; we should then have reason to suppose that here were masses of matter which, though not different in their origin from those that are gradually deposited at the bottom of the ocean, have been more acted upon by subterranean heat and the expanding power, that is to say, have been changed in a greater degree by the operations of the mineral region.

In that long sentence lies the discovery of metamorphic rock. But just as metamorphism will turn shale into slate, sandstone into quartzite, and granite into gneiss, Hutton had turned words into pumice. Unsurprisingly, his insights did not at once spread far and wide. They received a scattered following and much abuse. The attacks were theological, in the main, but, needless to say, geological as well—particularly with regard to his elastic sense of time. Even when people began to agree that the earth must be a great deal older than six thousand years, calculations were conservative and failed to yield the reach of time that Hutton's theory required. Lord Kelvin, as late as 1899, figured that twenty-five million years was the approximate age of the earth. Kelvin was the most august figure in contemporary science, and no one stepped up to argue. Hutton published his *Theory of the Earth* in 1795, when almost no one doubted the historical authenticity of Noah's Flood, and all species on earth were thought to have been created individually, each looking at the moment of its creation almost exactly as it did in modern times. Hutton disagreed with that, too. Writing a treatise on agriculture, he brought up the matter of variety in animals and noted, "In the infinite variation of the breed, that form best adapted to the exercise of the instinctive arts, by which the species is to live, will most certainly be continued in the propagation of this animal, and will be always tending more and more to perfect itself by the natural variation which is continually taking place. Thus, for example, where dogs are to live by the swiftness of their feet and the sharpness of their sight, the form best adapted to that end will be the most certain of remaining, while those forms that are less adapted to this manner of chase will be the first to perish; and, the same will hold with regard to all the other forms and faculties of the species, by which the instinctive arts of procuring its means of substance may be pursued." When he died, in 1797, Hutton was working on that manuscript, no part of which was published for a hundred and fifty years.

People who admired Hutton's theory of the earth became known—because of the theory's igneous aspects, its molten basalts and intruding granites—as vulcanists or plutonists, and they quickly grew to be the intellectual enemies of the Wernerian neptunists, and others who believed that God had made the world through a series of catastrophes, notably the Noachian flood. The schism between these two groups would carry well into the nineteenth and even into the twentieth century, the ratio gradually reversing. In 1800, the Huttonians were outnumbered at least ten to one. In fact, a Werner-trained neptunist took over the chair of natural history at the University of Edinburgh and for many years neptunism was official in Hutton's own city.

All this can be presumed to have bestirred John Playfair, a handsome, life-loving, and generous man of "mild majesty and considerate enthusiasm," as a contemporary described him. Never mind that the contemporary was his nephew. With all those neptunists and men of the cloth on the one side and his friend's prose on the other, the battle to Playfair must have seemed unjust, and he betook himself to alter the situation. The least of his many verbal gifts was a slow-cooled lucidity, a sense of the revealing phrase, and his *Illustrations of the Huttonian Theory of the Earth*, published in 1802, was the first fully clear and persuasive statement of what the theory was about. It is testimony to Playfair's efficacity that the opposition stiffened. "According to the conclusions of Dr. Hutton, and of many other geologists, our continents are of indefinite antiquity, they have been peopled we know not how, and mankind are wholly unacquainted with their origin," wrote the Calvinist geologist Jean André Deluc in 1809. "According to my conclusions, drawn from the same source, that of *facts*, our continents are of such small antiquity, that the memory of the revolution which gave them birth must still be preserved among men;

and thus we are led to seek in the book of Genesis the record of the history of the human race from its origin. Can any object of importance superior to this be found throughout the circle of natural science?"

As geologists built the time scale, their research and accumulating data imparted to Hutton's theory an obviously increasing glow. And in the early eighteen-thirties Charles Lyell, who said in so many words that his mission in geology was "freeing the science from Moses," gave Hutton's theory and his sense of deep time their largest advance toward universality. In three volumes, he published a work whose full title was *Principles of Geology, Being an Attempt to Explain the Former Changes of the Earth's Surface, by Reference to Causes Now in Operation.* Lyell was so anti-neptunist, so anti-catastrophist that he out-Huttoned Hutton both in manner and in form. He not only subscribed to the uniformitarian process—the topographical earth building and destroying and rebuilding itself through time—but was finicky in insisting that all processes had been going on at exactly the same rate through all ages. *Principles of Geology* was to be the most enduring and effective geological text ever published. The first volume was eighteen months off the press when H.M.S. Beagle set sail from Devonport with Charles Darwin aboard. "I had brought with me the first volume of Lyell's *Principles of Geology*, which I studied attentively; and the book was of the highest service to me in many ways. The very first place which I examined, namely St. Jago in the Cape de Verde islands, showed me clearly the wonderful superiority of Lyell's manner of treating geology, compared with that of any other author whose works I had with me or ever afterwards read." When Darwin had first studied geology, he had heard lectures in Wernerian neptunism at Edinburgh, and they had very nearly put him to sleep. Now he was voyaging on the Beagle and developing his own sense of the slow and repetitive cycles of the earth and the giddying depths of time, with Lyell's book in his hand and Hutton's theory in his head. In six thousand years, you could never grow wings on a reptile. With sixty million, however, you could have feathers, too.

According to present theory, many exotic terrains moved in from the western ocean and collected against North America during a span of nearly three hundred million years which ended roughly forty million years ago, increasing the continent to something like its present size. Three of these assembled at the latitude of Interstate 80. It was the first of these collisions that crunched and folded the wine-red sandstone near Carlin. The second, in the early Triassic, is what apparently caused the whole Carlin unconformity to revolve quite close to its present position. Sonomia, as the second terrain has been named, included much of what is now western Nevada and eastern California, and is said to have come into the continent with such force—notwithstanding that it was moving an inch or so a year—that it overlapped its predecessor by as much as eighty kilometres before it finally stopped. The evidence of this event is known locally as the Golconda Thrust, and both its upper and lower components are exposed in a big roadcut on the western flank of Golconda Summit, where the interstate, coming up out of Pumpernickel Valley, crosses a spur of the Sonoma Range. Small wonder that Deffeyes pulled over when we came to it and said, "Let's stick our eyeballs on this one."

It was dawn at the summit. We had been awake for hours and had eaten a roadhouse breakfast sitting by a window in which the interior of the room was reflected against the black of the morning outside while a television mounted on a wall behind us resounded with the hoofbeats of the great horse Silver. *The Lone Ranger.* Five A.M. CBS's good morning to Nevada. Waiting for bacon and eggs, I put two nickels in a slot machine and got two nickels back. The result was a certain radiance of mood. Deffeyes, for his part, was thinking today in troy ounces. It would take a whole lot more than two nickels to produce a similar effect on him. Out for silver, he was heading into the hills, but first, in his curiosity, he walked the interstate roadcut, now and again kicking a can. The November air was in frost. He seemed to be smoking his breath. He remarked that the mean distance between beer cans across the United States along I–80 seemed to be about one metre. Westward, tens of hundreds of square miles were etched out by the early light: basins, ranges, and—below us in the deep foreground—Paradise Valley, the village Golconda, sinuous stands of cottonwood at once marking and concealing the Humboldt. The whole country seemed to be steaming, vapors rising from warm ponds and hot springs. The roadcut was long, high, and benched. It was sandstone, for the most part, but at its lower, westernmost end the blasting had exposed a dark shale that had been much deformed and somewhat metamorphosed, the once even bedding now wrinkled and mashed—rock folded up like wet laundry. "You can spend hours doping out one of these shattered places, just milling around trying to find out what's going on," Deffeyes said cautiously, but he was fairly sure he knew what had happened, for the sandstone that lay above contained many volcanic fragments and was full of sharp-edged grains of chert and quartz, highly varied in texture, implying to him a volcanic source and swift deposition into the sea (almost no opportunity for streams to have rounded off the grains), implying, therefore, an island arc standing in deep water on a continental margin—an Aleutian chain, a Bismarck Archipelago, a Lesser Antilles, a New Zealand, a Japan, thrust upon and overlapping the established continent, a piece of which was that mashed-up shale. Deffeyes mused his way along the cut. "There is complexity here because you have not only the upper and lower plates of the Golconda Thrust, which happened in the early Triassic; you also have basin-range faulting scarcely a hundred yards away—enormously complicating the regional picture. If you look at a geologic map of western Canada and Alaska, you can see the distinct bands of terrain that successively attached themselves to the continent. Here the pattern has been all broken up and obscured by the block faulting of the Basin and Range, not to mention the great outpouring of Oligocene welded tuff. So this place is a handsome mess. If you ever want to study this sort of collision more straightforwardly, go to the Alps, where you had a continent-to-continent collision and that was it."

So much for theory. This roadcut contained both extremities of Deffeyes' wide interests in geology, and his attention was now drawn to a large gap in the sandstone, faulted open probably six or seven million years ago and now filled with rock crumbs, as if a bomb had gone off there in the ground. The material was gradated outward from a very obvious core. In a country full of living hot springs, this was a dead one. Sectioned by the road builders, it remembered in its swirls and convolutions the violence of water raging hot in rock. The dead hot spring had developed cracks, and they had been filled in by a couple of generations of calcite veins. Deffeyes was busy with his hammer, pinging, chipping samples of the calcite. "This stuff is too handsome to leave out here," he said, filling a canvas bag. "There was a lot of thermal action here. Most of this material is not even respectable rock anymore. It's like soil. In 1903, a mining geologist named Waldemar Lindgren found cinnabar in crud like this at Steamboat, near Reno. Cinnabar is mercury sulphide. He also found cinnabar in the fissures through which water had come up from deep in the crust. He thought, Aha! Mercury deposits are hot-spring deposits! And he applied that idea to ore deposits generally. He started classifying them according to the temperature of the water from which they were deposited—warm, hot, hotter, and so on. We know now that not all metal deposits are hydrothermal in origin, but more than half of them are. As you know, the hot water, circulating deep, picks up whatever is there—gold, silver, molybdenum, mercury, tin, uranium—and brings it up and precipitates it out near the surface. A vein of ore is the filling of a fissure. A map of former hot springs is remarkably close to a map of metal discoveries. Old hot springs like this one brought up the silver of Nevada. It would do my heart good to find silver right here in this roadcut and put it to the local highway engineer."

He took some samples, which eventually proved

to be innocent of silver, and we got back into the
pickup. We soon left the interstate for a secondary
road heading north—up a pastel valley, tan, with a
pale-green river course, fields of cattle and hay. It
was a valley that had been as special to the Paiutes
as the Black Hills were to the Sioux. The Paiutes
gave it up slowly, killing whites in desperation to
keep it, and thus bringing death on themselves. The
first pioneers to settle in this "desert" were farmers—
an indication of how lush and beautiful the basin
must have appeared to them, ten miles wide and
seventy miles long, framed in serrated ridges of
north-south-trending mountains: range, basin, range.
Magpies, looking like scale-model 747s, kept rising
into flight from the side of the road and gaining altitude over the hood of the pickup. Deffeyes said
they were underdeveloped and reminded him of
archaeopteryx, that first Jurassic bird. We crossed
cattle guards that were nothing more than stripes
painted on the road, indicating that Nevada cattle
may be underdeveloped, too, with I.Q.s in one digit,
slightly lower than the national norm.

For eight million years, Deffeyes was saying, as
the crustal blocks inexorably pulled apart here and
springs boiled up along the faults, silver had been
deposited throughout the Basin and Range. The continually growing mountains sometimes fractured
their own ore deposits, greatly complicating the se-

quence of events and confusing the picture for anyone who might come prospecting for ores. There was
another phenomenon, however, that had once made
prospecting dead simple. Erosion, breaking into hot-
spring and vein deposits, concentrated the silver.
Rainwater converted silver sulphides to silver
chloride, heavy stuff that stayed right where it was
and—through thousands of millennia—increased in
concentration as more rain fell. Such deposits, richer
than an Aztec dream, were known to geologists as
supergene enrichments. Miners called them surface
bonanzas. In the eighteen-sixties, and particularly in
the eighteen-seventies, they were discovered in range
after range. A big supergene enrichment might be
tens of yards wide and a mile long, lying at or near
the surface. Instant cities appeared beside them,
with false-front saloons and tent ghettos, houses of
sod, shanties made of barrels. The records of these
communities suggest uneven success in the settling
of disputes between partners over claims: "Davison
shot Butler through the left elbow, breaking the
bone, and in turn had one of his toes cut off with an
axe." They were places with names like Hardscrab-
ble, Gouge Eye, Battle Mountain, Treasure Hill. By
the eighteen-nineties, the boom was largely over and
gone. During those thirty years, there were more
communities in Nevada than there are now. "Silver is
our most depleted resource, because it gave itself

away," said Deffeyes, looking mournful. "You didn't
need a Ph.D. in geology to find a supergene enrichment."

All you needed was Silver Jim. Silver Jim was a
Paiute, and he, or a facsimile, took you up some valley or range and showed you grayish rock with
touches of green that had a dull waxy lustre like the
shine on the horn of a cow. Horn silver. It was just
lying there, difficult to lift. Silver Jim could show you
horn silver worth twenty-seven thousand dollars a
ton. Those were eighteen-sixties dollars and an un-
inflatable ton. You could fill a wheelbarrow and go
down the hill with five thousand dollars' worth of
silver. Three or four years ago, a miner friend of
Deffeyes who lives in Tombstone, Arizona, happened
to find on his own property an overlooked fragment
of a supergene enrichment, a narrow band no more
than a few inches thick, six feet below the cactus.
Knocking off some volcanic overburden with a front-
end loader, the miner went after this nineteenth-
century antique and fondly dug it out by hand. He
said to his children, "Pay attention to what I'm doing
here. Look closely at the rock. We will never see this
stuff again." In a couple of hours of a weekend afternoon, he took twenty thousand dollars from the
ground.

We were off on dirt roads now with a cone of
dust behind us, which Deffeyes characterized as the

local doorbell. He preferred not to ring it. This talkative and generous professor—who ordinarily shares his ideas as rapidly as they come to him, spilling them out in bunches like grapes—was narrow-eyed with secrecy today. He had stopped at a courthouse briefly, and—an antic figure, with his bagging sweater and his Beethoven hair—had revealed three digits to a county clerk in requesting to see a registry of claims. The claims were coded in six digits. Deffeyes kept the fourth, fifth, and sixth to himself like cards face down on a table. He found what he sought in the book of claims. Now, fifty miles up the valley, we had long since left behind us its only town, with its Odd Fellows Hall, its mercantile company, its cottonwoods and Lombardy poplars; and there were no houses, no structures, no cones of dust anywhere around us. The valley was narrowing. It ended where ranges joined. Some thousands of feet up the high face of a distant and treeless mountain we saw an unnaturally level line.

"Is that a road?" I asked him.

"That's where we're going," he said, and I wished he hadn't told me.

Looking up there, I took comfort in the reflection that I would scarcely be the first journalist to crawl out on a ledge in the hope of seeing someone else get rich. In 1869, the editor of the New York *Herald*, looking over his pool of available reporters, must have had no difficulty in choosing Tom Cash to report on supergene enrichments. Cash roved Nevada. He reported from one place that he took out his pocketknife and cut into the wall of a shaft, removing an ore of such obviously high assay that he could roll it in his fingers and it would not crumble. Cash told the mine owner that he feared being accused of exaggeration—"of making false statements, puffing"—with resulting damage to his journalistic reputation. There was a way to avoid this, he confided to the miner. "I would like to take a sample with me of some of the richest portions." The miner handed him a fourteen-pound rock containing about a hundred and fifty troy ounces of silver (seventy-three per cent). In the same year, Albert S. Evans, writing in the San Francisco *Alta California,* described a visit with a couple of bankers and a geologist to a claim in Nevada where he was lowered on a rope into a mine. "The light of our candles disclosed great black sparkling masses of silver on every side. The walls were of silver, the roof over our heads was of silver, and the very dust that filled our lungs and covered our boots and clothing with a gray coating was of fine silver. We were told that in this chamber a million dollars worth of silver lies exposed to the naked eye and our observations confirm the statement. How much lies back of it, Heaven only knows."

Heaven knew exactly. For while the supergene enrichments—in their prodigal dispersal through the Basin and Range—were some of the richest silver deposits ever discovered in the world, they were also the shallowest. There was just so much lying there, and it was truly bonanzan—to print money would take more time than to pick up this silver—but when it was gone it was gone, and it went quickly. Sometimes—as in the Comstock Lode in Virginia City—there were "true veins" in fissures below, containing silver of considerable value if more modest assay, but more often than not there was nothing below the enrichment. Mining and milling towns developed and died in less than a decade.

We were on our way to a nineteenth-century mine, and were now turning switchbacks and climbing the high mountainside. Deffeyes, in order to consult maps, had turned over the wheel to me. He said his interest in the secondary recovery of silver had been one result of certain computer models that had been given wide circulation in the early nineteen-seventies, using differential equations to link such things as world population, pollution, resources, and food, and allow them to swim forward through time, with a resulting prediction that the world was more or less going to come to an end by the year 2000, because it would run out of resources. "We have found all obvious deposits, and, true enough, we've got to pay the price," he said. "But they did not take into account reserves or future discoveries or picking over once again what the old-timers left behind." Seeking commissions from, for example, the Department of Energy, he began doing studies of expectable discoveries of oil and gas, of uranium. He sort of slid inadvertently from uranium into silver after a syndicate of New York businessmen came to him to ask for his help in their quest for gold. The group was called Eocene and was interested in scavenging old mines. Deffeyes pointed out to them that while new gold strikes are still occurring in the world and new gold mines are still being developed, no major silver mine has been discovered since 1915. The pressure for silver is immense. Dentistry and photography use two-thirds of what there is, and there are at present no commercial substitutes. "We've been wiped out. We've gone through it, just as we have gone through magnesium and bromine. You can raise the price of silver all you want to but you won't have a new mine." He predicted that as prices went up silver would probably outperform gold, as, by percentage, it overwhelmingly has done. The potentialities in the secondary recovery of silver appeared to him to be a lot more alluring than working through tailings for gold. Eocene engaged him as a consultant, to help them scavenge silver.

Now far above the basin, we were on the thin

line we had seen from below, a track no wider than the truck itself, crossing the face of the mountain. It curved into reentrants and out around noses and back into reentrants and out to more noses. I was on the inboard side, and every once in a while as we went around a nose I looked across the hood and saw nothing but sky—sky and the summits of a distant range. We could see sixty, seventy miles down the valley and three thousand feet down the mountain. The declivity was by no means sheer, just steep—a steepness, I judged, that would have caused the vehicle, had it slipped off the road, to go end over end enveloped in flame at a hundred yards a bounce. My hands slid on the wheel. They were filmed in their own grease.

The equanimous Deffeyes seemed to be enjoying the view. He said, "Where did you learn to drive a truck?"

"Not that it's so god-damned difficult," I told him, "but this is about the first time."

Before 1900, the method used in this country to extract silver from most ores was to stamp the rock to powder in small stamp mills, then stir the powder into hot salt water and mercury, and, after the mercury had attracted the silver, distill the mercury. In 1887, a more thorough extraction process had been developed in England whereby silver ores were dissolved in cyanide. The method moved quickly to

South Africa and eventually to the United States. An obvious application was to run cyanide through old tailings piles to see what others had missed, and a fair amount of such work was done, in particular during the Depression. There had been so many nineteenth-century mines in Nevada, however, that Deffeyes was sure that some had been ignored. He meant to look for them, and the first basin he prospected was the C Floor of Firestone Library, up the hill from his office in Princeton. There he ran through books and journals and began compiling a catalogue of mines and mills in the Basin and Range that had produced more than a certain number of dollars' worth of silver between 1860 and 1900. He prefers not to bandy the number. He found them in many places, from barrel-cactus country near the Colorado River to ranges near the Oregon line, from the Oquirrh Mountains of Utah to the eastern rampart of the Sierra Nevada. In all, he listed twenty-five. The larger ones, like the Comstock, had been worked and reworked and cyanided to death, and "tourists were all over them like ants." A scavenger had best consider lesser mines, out-of-the-way mines —the quick-shot enrichments, the small-fissure lodes, where towns grew and died in six years. He figured that any mine worth, say, a million dollars a hundred years ago would still be worth a million dollars, because the old mills at best extracted ninety per cent of the silver in the ores, and the ten per cent remaining would be worth about what the ninety per cent had been worth then. Pulling more books and journals off the shelves, he sought to learn if and where attention had been paid to various old mines in the nineteen-thirties, and wherever he discovered activity at that time he crossed off those mines.

His next move was to buy aerial photographs from the United States Geological Survey. The pictures were in overlapping pairs, and each pair covered sixteen square miles. "You look at them with stereo equipment and you are a giant with eyeballs a mile apart and forty thousand feet in the air. God, do you have stereovision! Things jump off the earth. You look for tailings. You look for dumps. You look for the faint scars of roads. The environmentalists are right. A scar in this climate will last. It takes a long time for the terrain to erase a road. You try to reason like a miner. If this was a mine, now where would I go for water? If this was a mill here, by this stream, then where is the mine? I was looking for mines that were not marked on maps. I could see dumps in some places. They stood out light gray. The old miners made dumps of rock that either contained no silver at all or did not contain enough silver to be worth their while at the time. I tried to guess roughly the volume of the dumps. Mill tailings made unnatural light-gray smudges on the pictures. Some of the tailings and dumps I found in these mountains appear on no maps I've seen."

He flew to Nevada, chartered a light plane, and went over the country a thousand feet above the ground, taking fresh private pictures with a telephoto lens. When he flew over places where other scavengers looked up and waved, he crossed those places off his list. He went in on the ground then, to a number of sites, and collected samples. He had machines at home that could deal with the samples in ways unheard of just a few years before, let alone in the nineteenth century. Kicking at old timbers, he looked at the nails. Wire nails came into use in 1900 and are convenient index fossils of the Age of Cyanide. He hoped for square nails.

Deffeyes was on his own now. His relationship with Eocene had faded out after they had chosen, on various points, to follow counsel other than his, and they transferred their scrutiny to Arizona, preferring not to cope with winter. One day in Princeton, his wife, Nancy Deffeyes, was looking through a stack of hundred-year-old *Engineering & Mining Journals* when she found a two-line reference to certain mining efforts in the eighteen-seventies that eventually assumed prominence on her husband's list, and that was what had brought him here and why we were crawling like a Japanese beetle across the face of this mountain.

We turned a last corner, with our inner wheels resting firmly on the road and the two others supported by Deffeyes' expectations. Now we were moving along one wall of a big V-shaped canyon that eventually became a gulch, a draw, a crease in the country, under cottonwoods. In the upper canyon, some hundreds of acres of very steep mountainside were filled with holes and shafts, hand-forged ore buckets, and old dry timbers. There were square nails in the timbers. An ore bucket was filled with square nails. "Good litter," Deffeyes said, and we walked uphill past the mine and along a small stream into the cottonwoods. The stream was nearly dry. Under the cottonwoods were the outlines of cabins almost a century gone. Here at seven thousand feet in this narrow mountain draw had lived a hundred people, who had held their last election a hundred years ago. They had a restaurant, a brewery, a bookstore. They had seven saloons. And now there was not so much as one dilapidated structure. There were only the old unhappy cottonwoods, looking alien and discontented over the moist bed of the creek. Sixteen stood there, twisted, surviving—most of them over four feet thick. "Those cottonwoods try an environmentalist's soul," Deffeyes said. "They transpire water like running fountains. If you were to cut them down, the creek would run. Cottonwoods drink the Humboldt. Some of the tension in this country is that miners need water. Getting rid of trees would preserve water. By the old brine-and-mercury method, it took three tons of water to mill one ton of ore. There was nothing like that in this creek. They had to take the ore from here to a big enough stream, and that, as it happens, was a twelve-mile journey using mules. They would have gone out of here with only the very best ore. There was probably a supergene enrichment here over a pretty good set of veins. They took what they took and were gone in six years."

We walked back down to the mine, below which the stream—in flash flood once or twice a century over several million years—had cut the deep sharp V of its remarkably plunging valley. A number of acres of one side had been used as a dump, and Deffeyes began to sample this unused ore. "They must have depended on what they could see in the rock," he said. "If it was easy to see, they got it all. If it was complicated and gradational, they couldn't differentiate as well, and I think they threw it here." The material was crumbly, loose, weathered, unstable underfoot, a pyramid side of decomposing shards. Filling small canvas bags at intervals of six feet, he worked his way across it. With each step, he sank in above his ankles. He was about two hundred feet above the stream. Given the steepness of the ground and the proximity of all the loose material to the critical angle of repose, I had no trouble imagining that he was about to avalanche, and that he would end up in an algal pool of the trickling stream below us, buried under megatons of unextracted silver. The little stream was a jumble of boulders, testimony of the floods, with phreatophytes around the boulders like implanted spears. Deffeyes obviously was happy and without a fear in the world. When a swift-rising wind blew dust in his face, he mooed. Working in cold sunshine with his orange-and-black conical cap on his head, he appeared to be the Gnome of Princeton, with evident ambition to escalate to Zurich.

To make a recovery operation worthwhile, he said, he would have to get five ounces of silver per ton. The figures would turn out to be better than that. Before long, he would have a little plastic-lined pond of weak cyanide, looked after by a couple of technicians, down where the ore from this mine had been milled. A blue streak in the tailings there would come in at fifty-eight ounces a ton—richer than any tailings he had ever found in Nevada. "You put cyanide on that ore, the silver leaps out of it," he would say. "I have enough cyanide there to kill Cincinnati. People have a love-hate relationship with cyanide. Abelson showed that lightning acts on carbon dioxide and other atmospheric components to make hydrogen cyanide, and hydrogen cyanide polymerizes and later reacts with water to form amino acids, which are the components of proteins—and that may be how life began. Phil Abelson is the editor of *Science*. He's a geochemist, and he worked on the Manhattan Project. To get the silver out of here at an acceptable price, you need small-scale technology. You need miniaturized equipment, simple techniques. In the nineteenth century, they made sagebrush fires to heat the brine to dissolve the silver chloride. When mercury picked up the silver, they knew they had 'the real stuff' from the squeak. A mercury-and-silver mixture is what the dentist uses, and when he mashes it into your tooth it makes the same squeak."

Deffeyes' methodology would depend on more than sagebrush and sound. In time, he would have a portable laboratory there, size of a two-hole privy, and in it would be, among other things, a silver single-ion electrode and an atomic-absorption spectrophotometer. He could turn on a flame, close two switches, and see at once the amount of silver in a sample. For a short while, he would have a five-pound ingot of raw silver on the floor, propping open the door. When he was finished with his pond, he would withdraw the cyanide and turn it into a marketable compound known as Prussian blue. He would cover his pond with dirt and sow it with crested wheat.

And now, finishing up his sampling at the mine in the mountains, he filled a large burlap bag with ore he would take home to improve his technique of extraction. The smaller samples he had taken were for assays of silver in various parts of the slope. "I'm nothing but a ragpicker," he said. "A scavenger armed with a forty-thousand-dollar X-ray machine." The wind picked up another cloud of dust off the dump and blew it into his face. He mooed. "That may feel like dirt to you, but it feels like money to me," he said.

"How much money would you say that felt like?" I asked him.

He took out a Magic Marker and began to do metric conversions, geometry, and arithmetic on the side of a new canvas bag. "Well, this section of the dump is at least fifteen thousand cubic metres," he said. "That is the most conservative figure. At two hundred dollars a ton, that works out to about three million dollars, left here in the side of the hill."

"What are those red stakes up there?"

"Somebody seems to think they're finding new ore. I'm interested in the old stuff, down here."

"If you've got good silver in those bags, what about Eocene? What if they decide they still own you? What if they go to the sheriff?"

"Eocene doesn't own me, and Eocene doesn't own the contents of my head. The law has long since decided that. But if anybody comes after me I want you to go to jail cheerfully rather than surrender your notes."

As we wound down the mountain at the end of the day, we stopped to regard the silent valley—the seventy miles of basin under a rouge sky, the circumvallate mountains, and, the better part of a hundred miles away, Sonoma Peak, of the Sonoma Range. Deffeyes said, "If you reduced the earth to the size of a baseball, you couldn't feel that mountain. With a telephoto lens, you could convince someone it was Everest." Even at this altitude, the air was scented powerfully with sage. There was coyote scat at our feet. In the dark, we drove back the way we had come, over the painted cattle guards and past jackrabbits dancing in the road, pitch-dark, and suddenly a Black Angus was there, standing broadside, middle of the road. With a scream of brakes, we stopped. The animal stood still, thinking, its eyes unmoving—a wall of beef. We moved slowly after that, and even more slowly when a white sphere materialized on our right in the moonless sky. It expanded some, like a cloud. Its light became so bright that we stopped finally and got out and looked up in awe. A smaller object, also spherical, moved out from within the large one, possibly from behind it. There was a Saturn-like ring around the smaller sphere. It moved here and there beside the large one for a few minutes and then went back inside. The story would be all over the papers the following day. The *Nevada State Journal* would describe a "Mysterious Ball of Light" that had been reported by various people at least a hundred miles in every direction from the place where we had been. "By this time we decided to get the hell out of there," a couple of hunters reported, "and hopped in our pickup and took off. As we looked back at it, we saw a smaller craft come out of the right lower corner. This smaller craft had a dome in the middle of it and two wings on either side, but the whole thing was oval-shaped." Someone else had said, "I thought it was an optical illusion at first, but it just kept coming closer and closer so that I could see it wasn't an illusion. Then something started coming out of the side of it. It looked like a star, and then a ring formed around it. A kind of ring like you'd see around Saturn. It didn't make any noises, and then it vanished."

"Now we're both believers," said one of the hunters. "And I don't ever want to see another one. We're pretty good-sized men and ain't scared of nothing except for snakes and now flying saucers."

After the small sphere disappeared, the large one rapidly faded and also disappeared. Deffeyes and I were left on the roadside among the starlighted eyes of dark and motionless cattle. "Copernicus took the world out of the center of the universe," he said. "Hutton took us out of a special place somewhere near the beginning of things and left us awash in the middle of the immensity of time. An extraterrestrial civilization could show us where we are with regard to the creation of life."

We also went to Jersey Valley, between the Fish Creek Mountains and the Tobin Range, where Deffeyes had once spent a couple of field seasons collecting data for his doctoral thesis. He had lived in a tent in the oven weather, and had chug-alugged water in quart draughts while examining the rising mountains and the sediments the mountains had shed. The thick welded tuff of the Oligocene catastrophe, having been the regional surface when the faulting started, was the first material to break into grains that washed and rolled downhill. When erosion wore through the tuff and into the older rock below, it sent the older rock also in fragments to the basins. Reading up through a basin was like reading down through a range. Deffeyes had locally described this record, and now he wished to relate its timing to the development of the province as a whole. Forty miles off the interstate and with a lot of dust settling behind, he paused on the brow of a small hill at the head of Jersey Valley. It was intimate, compared with others in the Basin and Range. For perhaps twenty miles, it ran on south between snow-covered mountains and was filled with a delirium of sage. Deffeyes let out a cowboy yell. There were cinder cones standing in the valley, young and basaltic—enormous black anthills of the Pleistocene. Here and there was a minor butte, an erosional remnant, kept intact by sandstone at the top, but approaching complete disintegration, and, like a melting sugar lump, soon to be absorbed into the basin plain. "In a lot of valleys in Nevada all you will see is sagebrush, and not know that eight feet below you is a hell of an interesting story," Deffeyes said. "I found late-Miocene horse teeth over there in the Tobin Range."

"How did you know they were late Miocene?"

"I didn't. I sent them to a horse-teeth expert. I also found beaver teeth, fish, a camel skeleton, and the jaw of a rhinoceros not so far from here. The jaw was late Miocene, too. Early- and middle-Miocene fossils are absent from the province. You'll remember that wherever we have found fossils in the basin sediments the oldest have been late Miocene. So, if the vertebrate paleontologists have their heads screwed on right, the beginning of the faulting of the Basin and Range can be dated to the late Miocene. Verte-brate paleontology is an important old sport, like tossing the caber."

We left the dirt road and drove a mile or so up a pair of ruts, then continued on foot across a rough cobbly slope. We went down into a dry gulch, climbed out of it, and walked along the contour of another slope. These declivities were not discrete hills but fragments of great alluvial fans that were spilling off the mountains and were creased by streams that were as dry as cracks in leather. In their intermittent way, these streams had exposed succes-sive layers of sediment, all of which happened to be dark in hue, with the pronounced exception of light-gray layers of ash. The ash was from elsewhere, from outside the province, punctuation brought in on the wind. It had come from volcanoes standing, prob-ably, in what is now the Snake River Plain, two hun-dred miles away. Settling into the basin lakes—long-gone Miocene, Pliocene lakes—much of it had turned into zeolites. Among the zeolites, Deffeyes had found in Jersey Valley three million tons of a variety called erionite, which is named for wool, and is fibrous, and when it gets into the linings of human lungs causes mesothelioma. There are more millions of tons of erionite throughout the Basin and Range, passively causing nothing. But if twenty-five valleys of the province were to be filled up with forty-six hundred concrete shelters for MX missiles, as the Defense Department has proposed, wind would present extraordinary hazards during the process of construction. It would be difficult to overestimate the amount of fine material that can be borne long distances by the wind. The largest single layer of ash that Deffeyes found in Jersey Valley was ten feet thick. He once showed it to Howel Williams, of the University of California at Berkeley, whom he regarded as "the greatest of volcanologists." Deffeyes asked Williams what might have been the size of a volcano that from two hundred miles away could send out such an explosion of ash. Williams just stood there impressed, shaking his head.

On a shelf above us was a pile of sticks of a size that in moister country could well have been collected by a beaver. "Hawk," Deffeyes said. "Note the southern exposure. The hawk went solar long ago. The sun incubates the eggs and the hawk is free to soar." Running his eye over the sequence of sediments revealed in the slope before us, he decided to begin his work right there. He was carrying in his hand a device of his own invention with which he hoped to accomplish the delicate operation of removing paleomagnetic samples from unconsolidated lake sediments. Less delicately, he had equipped me with a military shovel. He asked me to go along the slope digging foxholes a couple of feet deep in order to get rid of the weathered surface and prepare the way for him. As the mountains had given up grains and the grains had come down into the basin, any that had magnetite in them would have settled in a uniform manner, pointing like compasses toward the earth's magnetic pole. Since the late Miocene, the earth's magnetic field had reversed itself twenty times— from north to south, from south back to north—and the dates of those reversals had by now become well established. If Deffeyes could somehow collect un-consolidated but firmly compacted sediment and keep it from falling apart and destroying its own evidence while he carried it to a paleomagnetic lab, he might be able to compare what he already knew from his vertebrate time scale—his expertized horse teeth, his rhinoceros jaw—with the paleomagnetic time scale as expressed by the magnetite in the successive basin sediments. He would thus improve his knowledge of what occurred when—in this basin, this range. Later, he could correlate the ash falls and other stratigraphy of Jersey Valley with other valleys in the region, and make clearer the story of how it all took shape, adding polish to chapters of the Basin and Range. And so he had invented and machined a corer that would tap clear-plastic tubing gingerly into the earth with a micropiledriver made of non-magnetic aluminum. As I began the crude initial digging, Deffeyes said, "There are ten thousand feet of sediment here, and all of it has been deposited in eight million years. I have high hopes for the success of these endeavors. For each sample, I would prefer to go twenty feet into the slope instead of two. I would like to have a bulldozer as a substitute for you. But one has to settle for what one can get."

The first time I put my foot to the ground, the shovel broke in half. It was decapitated. After that, I had to hold its head in my hands and scrape as with an awkward trowel.

"There's more to this paleomagnetism game than reversals," Deffeyes said, "more than just determining when, and whether, the magnetic pole was in the north or south. The earth's magnetic field is such that a compass needle at the equator will lie flat, while a compass needle at the poles will want to stand straight up on end—with all possible gradations of that in the latitudes between. So by looking at the paleomagnetic compasses in rock you can tell not only whether the magnetic pole was in the north or south when the rock formed but also—from the more subtle positions of the needles—the latitude of the rock at the time it formed."

On the striated pavement of Algeria lies the till of polar glaciers. There are tropical atolls in Canada, tropical limestones in Siberia, tropical limestones in Antarctica. From fossils, from climates preserved in stone, such facts were known long before paleomagnetism was discovered; but they were, to say the least, imperfectly understood. Paleomagnetism, first perceived in 1906, eventually confirmed what the paleoclimatologists and paleontologists had been saying about the latitudes of origins of rocks, but it did not resolve the mystery of the phenomenon, because there seemed to be two equally reasonable explanations. Either the rock had moved (and continents with it) or the whole earth had rolled, like a child's top slowly turning on its side, and the poles and equator had wandered. Either the equator had gone to Minnesota or Minnesota had gone to the equator.

In an address to the German Geological Association in 1912, a meteorologist named Alfred Wegener suggested that the continents had moved. Wegener had received fame as a record-setting balloonist, an Arctic explorer, and now he was making an assertion for which his name would live in mockery for about fifty years. He discussed similarities among living creatures in lands separated by broad oceans. He discussed similarities in fossils found in separate parts of the world. He noted the obvious jigsaw relationship between landmasses that were separated by seas. He noted certain rock formations in Africa and South America that would fit right together if the Atlantic were closed. He knew nothing of paleomagnetism, which was in its infancy and was many years away from yielding insight to the problem, but he was the promulgator of the hypothesis of continental drift. Unfortunately, he attempted to explain how the continents moved. He envisioned them plowing like icebreakers through solid basalt. Almost no one believed his hypothesis, any more than they had believed Benjamin Franklin when, in 1782, possibly after a visit to Edinburgh, he said he thought that the surface parts of the earth were floating about on a liquid interior. Through the nineteen-thirties, and particularly after the Second World War, paleomagnetic data accrued, and, as it presented its story of kaleidoscopic environments changing through time in any given place, academic geologists sketched on globes and maps their curves of apparent polar wander. Here is where the poles were at the end of the Silurian; this is where they went from there. Rocks of identical age, sampled in various parts of the world, indicated as much in their imprisoned compasses.

Some geologists—little cells of them in South Africa, the odd don or two at Cambridge—preferred the other explanation, but they were few, and in geology departments around the world everybody would annually crowd in to hear Lucius P. Aenigmatite, Regius Professor of Historical Geology, give his world-renowned lecture ridiculing continental drift. Oil geologists, when they had found what they were looking for in deep sandstones put down by ancient rivers, naturally yearned to know in what direction those rivers had flowed. They had long since learned empirically that if you wanted to find the direction of the stream you had to use different pole positions for well cores of different ages. Whether this was the result of polar wander or continental drift did not much matter to the flying red horse. Other geologists satisfied themselves by deciding that the paleomagnetic compasses were unreliable, notwithstanding that oil companies were using them to make money. Certain English geologists produced confusion by embracing continental drift and then drawing up narratives and maps that showed continents moving all over the earth with respect to a fixed and undriftable England.

By the late nineteen-fifties, paleomagnetic evidence had piled up so high that it demanded improved explication. India, for example, yielded data that put it out of harmony with the rest of the world with respect to polar wander. Either there was an inexplicable series of anomalies in the data or India itself had moved, coming up from the Southern Hemisphere and completely crossing the equator, rapidly, and at a rate of speed (about ten centimetres a year) completely out of synchronization with the rate at which the equator's position had differed in other terrains. More data, and increased sophistication in the analysis of data, began to show that polar-wander curves—once thought to be in agreement worldwide—could differ some from continent to continent. Curves based on Paleozoic and Triassic rock in North America and in Europe looked much alike but, oddly, stood separate in the way that a single line will appear to be double in inebriate vision. The gap corresponds to the present width of the Atlantic Ocean. The opening of the Atlantic began in the Triassic.

If the hypothesis of continental drift had long been overshadowed by the hypothesis of polar wander, the reverse would before long be true. Researchers in paleomagnetism at Cambridge University concluded that their data were showing them that both hypotheses could be correct, as later research at Princeton would confirm. The poles indeed had wandered. The continents had moved as well. The phenomenon of "apparent polar wander" had been caused, right enough, by the movement of masses of land, but concomitantly the earth had rolled —and patterns of "true polar wander" were seen to be superimposed on all the other motions of the shifting surface of the world. But what motions? If the continents had drifted, then in what manner were they drifting? Where had they come from and where were they going? What would happen if two should collide? Since they obviously were not plowing through solid basalt, how in fact did they move?

It was all in a single decade—roughly 1957–67—that these questions were given answers of startling cohesion, as not only paleomagnetists but seismologists and oceanographers, geologists and geophysicists, whose specialties had been diverging through time, suddenly drew together around new outpourings of information and produced a chain of scientific papers whose interlocking insights would for most geologists fundamentally adjust their understanding of the dynamics of the earth.

"It was a change as profound as when we gave up the Biblical story," Deffeyes said as he tapped his collector into the ground. "It was a change as profound as Darwinian evolution, or Newtonian or Einsteinian physics."

The papers themselves had straightforward scientific titles, some of which—perhaps only in the afterlight of their great effect—seem to resound with the magnitude of the subject: "History of Ocean Basins," "Rises, Trenches, Great Faults, and Crustal Blocks," "Sea-Floor Spreading and Continental Drift," "Seismology and the New Global Tectonics." From Toronto, Princeton, San Diego, New York, Cambridge (England), there were about twenty primal contributions which, taken together, can be said to have constituted the plate-tectonics revolution.

Now, so few years later, plate tectonics is widely taken for granted. When I was in high school, there was essentially no television in America, and four years later television had replaced flypaper. When I was in high school, in the nineteen-forties, the term "plate tectonics" did not exist—albeit there was one remarkably prescient paragraph in our physical-geology textbook about the motions and mechanisms of continental drift. Today, children in schoolrooms just assume that the story being taught them is as old as the hills, and was told by God himself to their teacher in 4004 B.C.

The story is that everything is moving, that the outlines of continents by and large have nothing to do with these motions, that "continental drift" is actually a misnomer, that only the world picture according to Marco Polo makes much sense in the old-time browns and greens and Rand McNally blues. The earth is at present divided into some twenty crustal segments called plates. Plate boundaries miscellaneously run through continents, around continents, along the edges of continents, and down the middle of oceans. The plates are thin and rigid, like pieces of eggshell. In miles, sixty deep by five thousand by eight thousand are the dimensions of the Pacific Plate. "Pacific Plate" is not synonymous with "Pacific Ocean," which wholly or partly covers ten other plates. There are virtually no landmasses associated with some plates—the Cocos Plate, the Nazca Plate. Some plates are almost entirely land—the Arabian Plate, the Iran Plate, the (heaven help us)

China Plate. (Jokes may be invisible to some geologists. Harry Hess, who in 1960 opened out the new story with his "History of Ocean Basins," began it with these words: "The birth of the oceans is a matter of conjecture, the subsequent history is obscure, and the present structure is just beginning to be understood. Fascinating speculation on these subjects has been plentiful, but not much of it predating the last decade holds water.") Certain major plates are about half covered with ocean—the South American Plate, the African Plate, the North American Plate. Australia and India are parts of the same plate. It is shaped like a boomerang, with the landmasses at either end. Continents in themselves are not drifting, are not cruise ships travelling the sea. Continents are high parts of plates. East-west, the North American Plate starts in the middle of the Atlantic Ocean and ends in San Francisco. West-east, the Eurasian Plate begins in the middle of the Atlantic Ocean and ends in the sea of Okhotsk.

It is the plates that move. They all move. They move in varying directions and at different speeds. The Adriatic Plate is moving north. The African Plate came up behind it and drove it into Europe—drove Italy like a nail into Europe—and thus created the Alps. The South American Plate is moving west. The Nazca Plate is moving east. The Antarctic Plate is spinning, like pan ice in a river.

As has happened only twice before in geology—

with Abraham Werner's neptunist system and James Hutton's *Theory of the Earth*—the theory of plate tectonics has assembled numerous disparate phenomena into a single narrative. Where plates separate, they produce oceans. Where they collide, they make mountains. As oceans grow, and the two sides move apart, new seafloor comes into the middle. New seafloor is continuously forming at the trailing edge of the plate. Old seafloor, at the leading edge of a plate, dives into deep ocean trenches—the Kuril Trench, the Aleutian Trench, the Marianas Trench, the Java Trench, the Japan Trench, the Philippine Trench, the Peru-Chile Trench. The seafloor goes down four hundred miles after it goes into the trenches. On the way down, some of it melts, loses density, and—white-hot and violent—rises toward the surface of the earth, where it emerges as volcanoes, or stops below as stocks and batholiths, laccoliths and sills. Most of the volcanoes of the world are lined up behind the ocean trenches. Almost all earthquakes are movements of the boundaries of plates—shallow earthquakes at the trailing edges, where the plates are separating and new material is coming in, shallow earthquakes along the sides, where one plate is ruggedly sliding past another (the San Andreas Fault), and earthquakes of any depth down to four hundred miles below and beyond the trenches where plates are consumed (Alaska, 1964).

A seismologist discovered that deep earthquakes under a trench had occurred on a plane that was inclined forty-five degrees into the earth. As ocean floors reach trenches and move on down into the depths to be consumed, the average angle is something like that. Take a knife and cut into an orange at forty-five degrees. To cut straight down would be to produce a straight incision in the orange. If the blade is tilted forty-five degrees, the incision becomes an arc on the surface of the orange. If the knife blade melts inside, little volcanoes will come up through the pores of the skin, and together they will form arcs, island arcs—Japan, New Zealand, the Philippines, the New Hebrides, the Lesser Antilles, the Kurils, the Aleutians.

Where a trench happens to run along the edge of a continent and subducting seafloor dives under the land, the marginal terrain will rise. The two plates, pressing, will create mountains, and volcanoes will appear as well. The Peru-Chile Trench is right up against the west coast of South America. The Nazca Plate, moving east, is going down into the trench. Interspersed among the uplifted Andes are four thousand miles of volcanoes. The Pacific Ocean floor, going down to melt below that edge of the continent, has done much to help lift it twenty thousand feet.

Seafloor—ocean crust—is dense enough to go

down a trench, but continents are too light, too
buoyant. When a continent comes over a trench, it
will become stuck there, causing havoc. Australia is
such a continent, and where the Australian Plate has
jammed a trench it has buckled up the earth to make
the mountains of New Guinea, sixteen thousand five
hundred feet.

When two continental masses happen to move
on a collision course, they gradually close out the sea
between them—barging over trenches, shutting
them off—and when they hit they drive their leading
edges together as a high and sutured welt, resulting
in a new and larger continental mass. The Urals are
such a welt. So is the Himalaya. The Himalaya is the
crowning achievement of the vigorous Australian
Plate, of which India is the northernmost extremity.
India in the Oligocene, completing its long northward journey, crashed head on into Tibet, hit so hard
that it not only folded and buckled the plate
boundaries but also plowed in under the newly created Tibetan plateau and drove the Himalaya five
and a half miles into the sky. The mountains are in
some trouble. India has not stopped pushing them,
and they are still going up. Their height and volume
are already so great they are beginning to melt in
their own self-generated radioactive heat. When the
climbers in 1953 planted their flags on the highest
mountain, they set them in snow over the skeletons

of creatures that had lived in the warm clear ocean that India, moving north, blanked out. Possibly as much as twenty thousand feet below the seafloor, the skeletal remains had formed into rock. This one fact is a treatise in itself on the movements of the surface of the earth. If by some fiat I had to restrict all this writing to one sentence, this is the one I would choose: The summit of Mt. Everest is marine limestone.

Plates grow, shrink, combine, disappear, their number changing through time. They shift direction. Before the Pliocene, there was a trench off California. Seafloor moved into it from the west and dived eastward into the earth. Big volcanoes came up. Under the volcanoes, the melted crust cooled in huge volumes as new granite batholiths. Basin-range faulting has elevated the batholiths to fourteen thousand feet, and weather has sketched them out as the Sierra Nevada.

When seafloor goes into a trench, there can be a certain untidiness as segments are shaved off the top. They end up sitting on the other plate, large hunks of ocean crust that formed as much as a few thousand miles away and are now emplaced strangely among the formations of the continent. The California Coast Ranges—the hills of Vallejo, the hills of San Simeon, the hills of San Francisco—are a kind of berm that was pushed up out of the water by the incoming plate, including a jumble of oceanic and continental materials known to geologists as the Franciscan mélange. Geologists used to earn doctorates piecing together the stratigraphy of the Franciscan mélange, finding bedding planes in rock masses strewn here and there, and connecting them with dotted lines. Plate tectonics reveals that there is no stratigraphy in the mélange, no consecutive story of deposition—just mountains of bulldozed hash. The eastward motion of the ocean plate stopped soon after basin-range faulting began. The plate started moving in another direction. The trench, ceasing to be a trench then, was replaced by the San Andreas Fault.

Mountain building, in the Old Geology, had been seen as a series of orogenies rhythmically spaced through time, in part resulting from isostatic adjustments, and in part the work of "earth forces" that were not extensively explained. As mountains were disassembled, their materials were deposited in huge troughs, depressions, downbendings of the crust that were known as geosynclines. Earth forces made the geosynclines. As sediment accumulated in them, its weight pressed ever farther down into the mantle until the mantle would take no more, and then there came a trampoline effect, an isostatic bounce, that caused the material to rise. The Gulf of Mexico was a good example of a geosyncline, with a

large part of the Rocky Mountains sitting in it as
more than twenty-five thousand feet of silt, sand, and
mud, siltstone, sandstone, and shale. "The South will
rise again!" Deffeyes used to say. The huge body of
sediment would one day be lifted far above sea level
and dissected by weather and wrinkled into mountains in the way that the skin of an apple wrinkles as
the apple grows old and dry. The steady rhythm of
these orogenies was known as "the symphony of the
earth"—the Avalonian Orogeny in latest Precambrian time, the Taconic Orogeny in late Ordovician
time, the Acadian Orogeny in late Devonian time,
the Antler Orogeny in Mississippian time, the Alleghenian Orogeny in Permian time, the Laramide
Orogeny in Cretaceous-Tertiary time. It was a slow
march of global uplifting effects—predictable—
proceeding through history in stately order. By the
end of the nineteen-sixties, the symphony had come
to the last groove, and was up in the attic with the
old Aeolian. Mountain building had become a story
of random collisions, unpredictable, whims of the motions of the plates, which, when continents collided
or trenches otherwise jammed, could give up going
one way and move in another. The Avalonian, Taconic, Acadian, and Alleghenian Orogenies were
now seen, in plate theory, not as distinct events but
as successive parts of the same event, which involved
the closing of an ocean called Iapetus that existed

more or less where the Atlantic is today. As every scholar knows, Iapetus was the father of Atlas, for whom the Atlantic is named. The continents on either side of Iapetus came together not head on but like scissors closing from the north, folding and faulting their conjoining boundaries to make the Atlas Mountains and the Appalachian chain. It was a Paleozoic story, and the motions finally stopped. In the Mesozoic, an entirely new dynamic developed and the crust in the same region began to pull apart, to break into blocks that formed a new province, a Eurafrican-American basin and range. The blocks kept on separating until a new plate boundary formed, and eventually a new marine basin, which looked for a while like the Red Sea before widening to become an ocean.

It was on the mechanics of the seafloor that geology's revolutionary inquiries were primarily focussed in the early days. Harry Hess, a mineralogist who taught at Princeton, was the skipper of an attack transport during the Second World War, and he carried troops to landings—against the furious defenses of Iwo Jima, for example, and through rockets off the beaches of Lingayen Gulf. Loud noises above the surface scarcely distracted him. He had brought along a new kind of instrument called a Fathometer, and, battle or no battle, he never turned it off. Its stylus was drawing pictures of the floor of the sea. Among the many things he discerned there were dead volcanoes, spread out around the Pacific bottom like Hershey's Kisses on a tray. They had the arresting feature that their tops had been cut off, evidently the work of waves. Most of them were covered with thousands of feet of water. He did not know what to make of them. He named them guyots, for a nineteenth-century geologist at Princeton, and sailed on.

The Second World War was a technological piñata, and, with their new Fathometers and proton-precession magnetometers, oceanographers of the nineteen-fifties—most notably Bruce Heezen and Marie Tharp at Columbia University—mapped the seafloor in such extraordinary detail that in a sense they were seeing it for the first time. (Today, the very best maps are classified, because they reveal the places where submarines hide.) What stood out even more prominently than the deep trenches were mountain ranges that rose some six thousand feet above the general seafloor and ran like seams through every ocean and all around the globe. They became known as rises, or ridges—the Mid-Atlantic Ridge, the Southeast Indian Ocean Ridge, the East Pacific Rise. They fell away gently from their central ridgelines, and the slopes extended outward hundreds of miles, to the edges of abyssal plains—the Hatteras Abyssal Plain, the Demerara Abyssal Plain, the Tasman Abyssal Plain. Right down the spines of most of the submarine cordilleras ran high axial valleys, grooves that marked the summit line. These eventually came to be regarded as rift valleys, for they proved to be the boundaries between separating plates. As early as 1956, oceanographers at Columbia had assembled seismological data suggesting that a remarkable percentage of all earthquakes were occurring in the mid-ocean rifts—a finding that was supported, and then some, after a worldwide system of more than a hundred seismological monitoring stations was established in anticipation of the nuclear-test-ban treaty of 1963. If there was to be underground testing, one had to be able to detect someone else's tests, so a by-product of the Cold War was seismological data on a scale unapproached before. The whole of plate tectonics, a story of steady-state violence along boundaries, was being brought to light largely as a result of the development of instruments of war. Earthquakes "focus" where earth begins to move, and along transform faults like the San Andreas the focusses were shallow. At the ocean trenches they could be very deep. The facts accrued. Global maps of the new seismological data showed earthquakes not only clustered like stitchings all along the ridges of the seafloor mountains but also in the trenches and transform faults, with the result that the seismology was sketching the earth's crustal plates.

To Rear Admiral Hess, as he had become in the U.S. Naval Reserve, it now seemed apparent that seafloors were spreading away from mid-ocean ridges, where new seafloor was continuously being created in deep cracks, and, thinking through as many related phenomena as he was able to discern at the time, he marshalled his own research and the published work of others up to 1960 and wrote in that year his "History of Ocean Basins." In the nineteen-forties, a professor at Delft had written a book called *The Pulse of the Earth*, in which he asserted with mild cynicism that where gaps exist among the facts of geology the space between is often filled with things "geopoetical," and now Hess, with good-humored candor, adopted the term and announced in his first paragraph that while he meant "not to travel any further into the realm of fantasy than is absolutely necessary," he nonetheless looked upon what he was about to present as "an essay in geopoetry." He could not be sure which of his suppositions might be empty conjecture and which might in retrospect be regarded as precocious insights. His criterion could only have been that they seemed compelling to him. His guyots, he had by now decided, were volcanoes that grew at spreading centers, where they protruded above the ocean surface and were attacked by waves. With the moving ocean floor they travelled slowly down to the abyssal plains and went on eventually to "ride down into the jaw crusher" of the deep trenches, where they were consumed. "The earth is a dynamic body with its surface constantly changing," he wrote, and he agreed with others that the force driving it all must be heat from deep in the mantle, moving in huge revolving cells (an idea that had been around in one form or another since 1839 and is still the prevailing guess in answer to the unresolved question: What is the engine of plate tectonics?). Hess reasoned also that the heat involved in the making of new seafloor is what keeps the ocean rises high, and that moving outward the new material gradually cools and subsides. The rises seemed to be impermanent features, the seafloor altogether "ephemeral." "The whole ocean is virtually swept clean (replaced by new mantle material) every three hundred to four hundred million years," he wrote. "This accounts for the relatively thin veneer of sediments on the ocean floor, the relatively small number of volcanic seamounts, and the present absence of evidence of rocks older than Cretaceous in the oceans." In ending, he said, "The writer has attempted to invent an evolution for ocean basins. It is hardly likely that all of the numerous assumptions made are correct. Nevertheless it appears to be a useful framework for testing various and sundry groups of hypotheses relating to the oceans. It is hoped that the framework with necessary patching and repair may eventually form the basis for a new and sounder structure."

In 1963, Drummond Matthews and Fred Vine,

of Cambridge University, published an extraordinary piece of science that gave to Hess's structure much added strength. Magnetometers dragged back and forth across the seas had recorded magnetism of two quite different intensities. Plotted on a map, these magnetic differences ran in stripes that were parallel to the mid-ocean ridges. The magnetism over the centers of the ridges themselves was uniformly strong. Moving away from the ridges, the strong and weak stripes varied in width from a few kilometres to as many as eighty. Vine and Matthews, chatting over tea in Cambridge, thought of using this data to connect Harry Hess's spreading seafloor to the time scale of paleomagnetic reversals. The match would turn out to be exact. The weaker stripes matched times when the earth's magnetic field had been reversed, and the strong ones matched times when the magnetic pole was in the north. Moreover, the two sets of stripes—calendars, in effect, moving away from the ridge—seemed to be symmetrical. The seafloor was not only spreading. It was documenting its age. L. W. Morley, a Canadian, independently had reached the same conclusions. Vine and Matthews' paper was published in the British journal *Nature* in September, 1963, and became salient in the development of plate tectonics. In January of the same year, Morley had submitted almost identical ideas to the editors of *Nature*, but they were not yet prepared to accept them, so Morley then submitted the paper in the United States to the *Journal of Geophysical Research*, which rejected it summarily. Morley's paper came back with a note telling him that his ideas were suitable for a cocktail party but not for a serious publication.

Data confirming the Vine-Matthews hypothesis began to accumulate, nowhere more emphatically than in a magnetic profile of the seafloor made by the National Science Foundation's ship Eltanin crossing the East Pacific Rise. The Eltanin's data showed that the seafloor became older and older with distance from the spreading center, and with perfect symmetry for two thousand kilometres on either side. All through the nineteen-sixties, ships continued to cruise the oceans dragging magnetometers behind, and eventually computers were programmed to correlate the benthic data with the surface wanderings of the ships. Potassium-argon dating had timed the earth's magnetic reversals to apparent perfection for the last three and a half million years. Geologists at Columbia calculated the rate of seafloor spreading for those years and then assumed the rate to have been constant through earlier time. On that assumption, they extrapolated a much more extensive paleomagnetic time scale. (Improved radiometric dating later endorsed the accuracy of the method.) And with that scale they swiftly mapped the history of ocean basins. Compared with a geologic map of a continent, it was a picture handsome and spare. As the paleomagnetist Allan Cox, of Stanford University, would describe it in a book called *Plate Tectonics and Geomagnetic Reversals*, "The structure of the seafloor is as simple as a set of tree rings, and like a modern bank check it carries an easily decipherable magnetic signature."

Meanwhile, geophysicists at Toronto, Columbia, Princeton, and the Scripps Institution of Oceanography were filling in the last major components of the plate-tectonics paradigm. They figured out the geometry of moving segments on a sphere, showed that deformation happens only at the margins of plates, charted the relative motions of the plates, and mapped for the first time the plate boundaries of the world.

If it was altogether true, as Hess had claimed, that with relative frequency "the whole ocean is virtually swept clean," then old rock should be absent from deep ocean floors. Since 1968, the drill ship Glomar Challenger has travelled the world looking for, among other things, the oldest ocean rocks. The oldest ever found is Jurassic. In a world that is 4.6 billion years old, with continental-shield rock that has been dated to 3.8, it is indeed astonishing that the oldest rock that human beings have ever removed from a seafloor has an age of a hundred and fifty million years—that the earth is thirty times as old as the oldest rock of the oceans. In 1969, it seemed likely that the oldest ocean floor would be found in the Northwest Pacific. The Glomar Challenger went there to see. Two Russians were aboard who believed that rock older than Jurassic—rock of the Paleozoic, in all likelihood—would be discovered. They took vodka with them to toast the first trilobite to appear on deck. Trilobites, index fossils of the Paleozoic, came into the world at the base of the Cambrian and went out forever in the Permian Extinction—eighty million years before the age of the oldest rock ever found in modern oceans. As expected, the oceanic basement became older and older as the ship drilled westward from Hawaii. But even at the edge of the Marianas Trench, the Russians were disappointed. No vodka. Ah, but there might be older rock on the other side of the trench, in the floor of the Philippine Sea. The ship pulled up its drill pipe and moved across the trench. This time the rock was Miocene, more or less a tenth as old as the Jurassic floor. The Russians broke out the vodka. A toast! Neil Armstrong and Edwin Aldrin were walking around on the moon.

"In the old days we would have called this North America," Deffeyes said, sinking another clear tube into the ground. "We now think of plates. The plate-tectonics revolution came as a surprise, with very little buildup to it. There was none of that cloud that precedes a political revolution. In the nineteen-fifties, when I was a graduate student, nearly all the faculty at Princeton thought continental drift was sheer baloney. A couple of years later, Harry Hess broke it open. I had thought I would go through my career without anything like it. Oil and mining seemed enough of a contribution to keep one going. But now something had come along that was so profound that it took the whole science with it. We used to think that continents grew like onions around old rock. That was overturned by plate tec-

tonics. And we could see now how amazingly fast
you could put up a mountain range. A continent-to-
continent collision was a hell of an episode at a lim-
ited place. After the Appalachians and the Urals
were recognized as continent-to-continent sutures,
people said, 'O.K., where's the suture in California?'
Geologists kept jumping up and saying, 'I've got the
suture! I've got the suture!' It turned out, of course,
that there were at least three sutures. In each instance, a great island had closed up a sea and hit into
America—just as India hit Tibet, just as Kodiak Island, which is a mini-India, is about to plow into
Alaska. Fossils from the mid-Pacific have been found
here in the West, and limestones that lithified a
thousand miles south of the equator. Formations in
California have alien fossils with cousins in the rock
of New Guinea. For a while, people were going
around naming a defunct ocean for every suture. The
first piece, coming in from the west, was the one that
rode up onto North America about forty miles, not a
trivial distance, in Mississippian time. That was the
action that first tipped the rock in the Carlin unconformity. The old name for it was the Antler Orogeny.
In the Triassic, the second one arrived—the Golconda Thrust—and rode fifty miles over the trailing
edge of the first one, and in the Jurassic the third one
came in, sutured on somewhere near Sacramento,
and more or less completed California. I have read

that two geologists have found in Siberia a displaced terrain that was taken off of North America. The Lord giveth and the Lord taketh away."

I mentioned that I had read in *Geology* that one out of eight geologists does not accept plate-tectonic theory.

He said, "There are still a few people dragging their feet. They don't want to come into the story."

I asked him if he thought the Uinta Mountains could be explained in terms of plate theory. The Uintas are a range in the Rockies, seven hundred miles from the sea, and they run east-west, unlike virtually all other ranges for thousands of miles around them. If the western cordilleras were raised by colliding plates, how did the Uintas happen to come up at right angles to the other mountains?

He said, "You must have been talking to a Rocky Mountain geologist." He said nothing else for a time, while he tapped at the earth I had uncovered and captured a perfect sample. Then he said, "The north side of the Uintas is a spectacular mountain wall. Glorious. You come upon it and suddenly you see structurally the boundary of the range. But you don't see what put it there. The Uintas are mysterious. They are not a basin-range fault block, yet they have come up nearly vertically, with almost no compression evident. You just stand there and watch them go up into the sky. They don't fit our idea of plate tectonics. The Rockies in general will be one of the last places in the world to be deciphered in terms of how many hits created them, and just when, and from where."

The article in *Geology* was based on a questionnaire that was circulated toward the end of the nineteen-seventies. The results indicated that forty per cent of geologists had come to feel that plate theory was "essentially established," while a roughly equal number preferred to qualify a bit and say that it was "fairly well established." Eleven per cent felt that the theory was "inadequately proven." Seven per cent said they had accepted continental drift before 1940. Six per cent thought plate tectonics would be "still in doubt" in the late nineteen-eighties. And one geologist predicted that the theory would eventually be rejected.

"At any given moment, no two geologists are going to have in their heads exactly the same levels of acceptance of all hypotheses and theories that are floating around," Deffeyes said. "There are always many ideas in various stages of acceptance. That is how science works. Ideas range from the solidly accepted to the literally half-baked—those in the process of forming, the sorts of things about which people call each other up in the middle of the night. All science involves speculation, and few sciences include as much speculation as geology. Is the Delaware Water Gap the outlet of a huge lake all other traces of which have since disappeared? A geomorphologist will tell you that, in principle, the idea is O.K. You have to deal with partial information. In oil drilling, you had better be ready to act shrewdly on partial information. Do physicists do that? Hell, no. They want to have it to seven decimal places on their Hewlett-Packards. The geologist has to choose the course of action with the best statistical chance. As a result, the style of geology is full of inferences, and they change. No one has ever seen a geosyncline. No one has ever seen the welding of tuff. No one has ever seen a granite batholith intrude."

Since I was digging his sample pits, I felt enfranchised to remark on what I took to be the literary timbre of his science.

"There's an essential difference," he said. "The authors of literary works may not have intended all the subtleties, complexities, undertones, and overtones that are attributed to them by critics and by students writing doctoral theses."

"That is what God says about geologists," I told him, chipping into the sediment with his broken shovel.

"You may recall Archelaus's explanation of earthquakes," he said cryptically. "Earthquakes were caused by air trapped in underground caves. It shook the earth in its effort to escape. Everyone knew then that the earth was flatulent."

Deffeyes said he had asked his friend Jason Morgan—whose paper "Rises, Trenches, Great Faults, and Crustal Blocks" defined the boundaries and motions of the plates—what he was going to do for an encore. Morgan said he didn't know, but possibly the most exciting thing to do next would be to prove the theory wrong.

That would be a reversal comparable to the debunking of Genesis. I remembered Eldridge Moores, of the University of California at Davis, telling me what it had been like to be in graduate school at the height of the plate-tectonics revolution, and how he had imagined that the fervor and causal excitement of it was something like landing on Guadalcanal in the middle of the action of "a noble war." Tanya Atwater, a marine geologist who eventually joined the faculty of the Massachusetts Institute of Technology, was then a graduate student at the Scripps Institution of Oceanography. In a letter written to Allan Cox at Stanford, she re-creates the milieu of the time. "Seafloor spreading was a wonderful concept because it could explain so much of what we knew, but plate tectonics really set us free and flying. It gave us some firm rules so that we could predict what we should find in unknown places. . . . From the moment the plate concept was introduced, the geometry of the San Andreas system was an obviously interesting example. The night Dan McKenzie and Bob Parker told me the idea, a bunch of us were drinking beer at the Little Bavaria in La Jolla. Dan sketched it on a napkin. 'Aha!' said I, 'but what about the Mendocino trend?' 'Easy!,' and he showed me three plates. As simple as that! The simplicity and power of the geometry of those three plates captured my mind that night and has never let go since. It is a wondrous thing to have the random facts in one's head suddenly fall into the slots of an orderly framework. It is like an explosion inside. That is what happened to me that night and that is what I often felt happen to me and to others as I was working out (and talking out) the geometry of the western U.S.... The best part of the plate business is that it has made us all start communicating. People who squeeze rocks and people who identify deep-ocean nannofossils and people who map faults in Montana suddenly all care about each other's work. I think I spend half my time just talking and listening to people from many fields, searching together for how it might all fit together. And when something does fall into place, there is that mental explosion and the wondrous excitement. I think the human brain must love order."

Deffeyes, meanwhile, had joined Shell before the excitement developed. Growing up in oil fields, he had grown to like them, to admire the skill and independence of the crews, the competent manner in which they lived with danger. "Like a bullfighter, you are careful. So danger is not an overwhelming risk. But it is always there. And you can be crushed, burned, asphyxiated, destroyed by an explosion. A crew on a rig floor runs pipe in the hole with swift precision, and any piece of equipment can take your hand off just as fast." As a small boy, he often went into oil fields with his father, whose assignments changed many times—Oklahoma City, Hutchinson, Great Bend, Midland, Hobbs, Casper. As a teen-ager, Deffeyes played the French horn in the Casper Civic Symphony. He debated on the high-school team. He became—as he has remained—a forensic marvel, the final syllables of his participles and gerunds ringing like Buddha's gongs. In the way that others collected stamps, he collected rocks. For counsel, he took his specimens to the geologists in town, of whom there were plenty, including Paul Walton, who in 1948 had suggested to J. Paul Getty that he go to Kuwait. High-school summers, Deffeyes worked as an assistant shooter with seismic crews and as a roustabout maintaining wells. When he finished graduate school and moved on to Houston with Shell, he was ignorant not only of the imminent revolution in geologic theory but also of the approaching atrophy in successful exploration for oil. M. King Hubbert, an outstanding geological geophysicist, was with Shell at the time, and Deffeyes had only settled in when Hubbert happened to predict (with amazing accuracy) the approaching date when more oil would be coming out of American ground than geologists would be discovering. He predicted the energy crisis that would inevitably follow. When Deffeyes saw Hubbert's evanescing figures, he saw disappearing with them what had looked to be his most productive years. A decade before the first gas lines, he resigned from Shell to go into teaching and was soon on the faculty at Oregon State, where he set himself up as a chemical oceanographer, because the ocean was where things were happening. The university had bought from the government a small ship left over from the Second World War and had converted it for oceanographic research. "Working for an oil company had suddenly become like working for a railroad—a dying industry. Now in this new field new equipment was being improvised, and the problems were the same as they were in the oil fields when I was a kid. In the ocean, we used bottom-hole pressure gauges and other oil-field equipment. I could feel the same sort of excitement I had felt years before in the oil fields, and with the same sorts of people—roustabouts and roughnecks—in the crew."

Unfortunately, Deffeyes had a signal defect as an oceanographer. He got terribly seasick. His enthusiasm grew moist, and he began to contrive to remain ashore. Then in October, 1965, J. Tuzo Wil-

son, of the University of Toronto, and Fred Vine, of
Cambridge, published a paper in which they defined
an oddly isolated piece of mid-ocean ridge off the
coasts of Oregon and Washington. It was the spreading center of what would eventually become known
as the Juan de Fuca Plate, one of the smallest of all
the crustal plates in the world. The volcanoes of the
Cascades—Mt. Hood, Mt. Rainier, Mt. St. Helens,
Glacier Peak—were lined up behind its trench.
"Continental drift is one hypothesis I'll get seasick
for," Deffeyes decided, and he signed up for a week's
use of the ship. He had no program in mind. To ask
for a suggestion, he picked up a telephone and called
Harry Hess. Instantly, Hess said, "Go to the ridge
and dredge some rock from the axial valley. It better
not be old." Hess's hypothesis that new seafloor
forms at ocean rises had scarcely been tested. This
was before the Eltanin profile, and before the voy-
ages of the Glomar Challenger. Hess's immediate
response to Deffeyes was to suggest a test that could
have shelved his hypothesis then and there.

Deffeyes went out to dredge the rock, but first
he had to find the ridge, so he made a long pass with
his echo sounder tracing the profile of the bottom.
The ridge-axis rock, when he dredged it up, was extremely young. But what in the end interested
Deffeyes at least as much was the benthic profile that
had been traced by the stylus of the sounder. The

profile of the spreading center in the ocean bottom off Oregon seemed remarkably familiar to someone who had done his thesis field work in Nevada. It appeared to be, in miniature, a cross-section of the Basin and Range. The new crust, spreading out, had broken into fault blocks and had become a microcosm of the Basin and Range, because both were expressions of the same cause. It was a microcosm, too, of the Triassic lowlands of the East two hundred million years ago—Triassic Connecticut, Triassic New Jersey—with their border faults and basalt flows, their basins and ranges, gradually extending, pulling apart, to open the Atlantic. The Red Sea of today was what the Atlantic and its two sides had looked like about five million years after the Atlantic began to open. The Red Sea today was what the Basin and Range would probably look like at some time in the future.

In December, 1972, the astronaut Harrison Schmitt, riding in Apollo 17, looked down at the Red Sea and the Gulf of Aden—at a simple geometry that seemed to have been made with a jigsaw barely separating Africa and the Arabian peninsula. He told the people at Mission Control, "I didn't grow up with the idea of drifting continents and seafloor spreading. But I tell you, when you look at the way pieces of the northeastern portion of the African continent seem to fit together, separated by a narrow gulf, you could make a believer out of anybody." Schmitt was one of the eighty per cent who were changing their minds about the new global theory. In addition to his astronaut's training, he had a Ph.D. in geology, and he would bring back from the moon a hundred kilos of rock.

Twenty miles out of Winnemucca, and the interstate is dropping south toward the Humboldt Range. A coyote runs along beside the road. It is out of its element, tongue out, outclassed, under minimum speed. Deffeyes says that most ranges in the Basin and Range had one or two silver deposits in them, if any, but the Humboldts had five. We have also entered the bottomlands of the former Lake Lahontan. The hot-springs map shows more activity in this part of the province. Extension of the earth's crust has been somewhat more pronounced here, Deffeyes explains, and hence there are more ore deposits. He feels that when a seaway opens up, the spreading center will be somewhere nearby. Or possibly back in Utah, in the bed of Lake Bonneville. "But this one has better connections."

"Connections?"

"Death Valley. Walker Lake. Carson Sink." An Exxon map of the western United States is spread open on the seat between us. He runs his finger from Death Valley to Carson Sink and on northward to cross the interstate at Lovelock. "The ocean will open here," he repeats. "Or in the Bonneville basin. I think here."

A few miles off the road is the site of a planned community dating from the nineteen-sixties. It was to have wide streets and a fountained square, but construction was delayed and then indefinitely postponed. Ghostless ghost town, it had been named Neptune City.

With the river on our right, we round the nose of the Humboldt Range, as did the Donner Party and roughly a hundred and sixty-five thousand other people, in a seventeen-year period, heading in their wagons toward Humboldt Sink, Carson Sink, and the terror of days without water. But first, as we do now, they came into broad green flats abundantly fertile with grass, knee-high grass, a fill for the oxen, the last gesture of the river before it vanished into the air. The emigrants called this place the Big Meadows of the Humboldt, and something like two hundred and fifty wagons would be resting here at any given time.

"There was a sea here in the Triassic," Deffeyes

remarks. "At least until the Sonomia terrain came in and sutured on. The sea was full of pelagic squid, and was not abyssal, but it was deep enough so the bottom received no sunlight, and bottom life was not dominant."

"How do you know it was not dominant?"

"Because I have looked at the siltstones and the ammonites in them, and that is what I see there."

Visions of oceans before and behind us in time, we roll on into Lovelock. SLOW—DUST HAZARD. Lovelock, Nevada 89419. There are cumulus snow clouds overhead and big bays of blue in the cold sky, with snow coming down in curtains over the Trinity Range, snow pluming upward over the valley like smoke from a runaway fire. Lovelock was a station of the Overland Stage. It became known throughout Nevada as "a good town with a bad water supply." An editor of the Lovelock *Review-Miner* wrote in 1915, "There is little use in trying to induce people to locate here until the water question is settled. . . . Maybe the water does not kill anyone, but it certainly drives people away." In 1917, Lovelock was incorporated as a third-class city, and one of its first acts was to enforce a ban on houses of prostitution within twelve hundred feet of the Methodist Episcopal Church. Another was a curfew. Another ordered all city lights turned off when there was enough moon.

JAX CASINO LIBERAL SLOTS

·

TWO STIFFS SELLING GAS AND MOTEL

·

WATER SUPPLY FROM PRIVATE WELL

·

LOVELOCK SEED COMPANY

GRAINS AND FEED

Here in the Big Meadows of the Humboldt, the principal employer is the co-op seed mill on the edge of town, which sends alfalfa all over the world.

On the sidewalks are men in Stetsons, men in three-piece suits, men in windbreakers, tall gaunt overalled men with beards. There are women in Stetsons, boots, and jeans. A thin young man climbs out of a pickup that is painted in glossy swirls of yellow and purple, and has a roll bar, balloon tires, headphones, and seventeen lights.

There are terraces of Lake Lahontan above the ballfield of the Lovelock Mustangs. Cattle graze beside the field. The Ten Commandments are carved in a large piece of metamorphosed granite outside the county courthouse.

NO. 10:

THOU SHALT NOT COVET THY NEIGHBOR'S WIFE,

NOR HIS MANSERVANT, NOR HIS MAIDSERVANT,

NOR HIS CATTLE

·

BRAZEN ONAGER—BAR—BUD—PIZZA

·

WHOO-O-A MOTEL

"Lovelock was a person's name," Deffeyes cautions.

LOVELOCK MERCANTILE

The name is fading on the cornice of Lovelock Mercantile. It was built in 1905, expanded in 1907, is the bus stop now, liquor store, clothing store, grocery store, real-estate office, bakery, Western Union office —all in one room. There is a sign on one of the columns that hold up the room:

WE CANNOT ACCEPT

GOVERNMENT MEAL TICKETS

Across the valley is a huge whitewash "L" on a rock above the fault scar of the Humboldt Range.

We go into Sturgeon's Log Cabin restaurant and sit down for coffee against a backdrop of rolling cherries, watermelons, and bells. A mountain lion in a glass case. Six feet to the tip of the tail. Shot by Daniel (Bill) Milich, in the Tobin Range.

I hand Deffeyes the Exxon map and ask him to sketch in for me the opening of the new seaway, the spreading center as he sees it coming. "Of course, all the valleys in the Great Basin are to a greater or lesser extent competing," he says. "But I'd put it where I said—right here." With a pencil he begins to rough in a double line, a swath, about fifteen miles wide. He sketches it through the axis of Death Valley and up into Nevada, and then north by northwest through Basalt and Coaldale before bending due north through Walker Lake, Fallon, and Lovelock. "The spreading center would connect with a transform fault coming in from Cape Mendocino," he adds, and he sketches such a line from the California coast to a point a little north of Lovelock. He is sketching the creation of a crustal plate, and he seems confident of that edge, for the Mendocino transform fault—the Mendocino trend—is in place now, ready to go. He is less certain about the southern edge of the new plate, because he has two choices. The Garlock Fault runs east-west just above Los Angeles, and that could become a side of the new plate; or the spreading center could continue south through the Mojave Desert and the Salton Sea to meet the Pacific Plate in the Gulf of California. "The Mojave sits in there with discontinued basin-and-range faulting," Deffeyes says, almost to himself, a substitute for whistling, as he sketches in the alternative lines. "There has to be a transform fault at the south end of the live, expanding rift. The sea has got to get through somewhere."

Now he places his hands on the map so that they frame the Garlock and Mendocino faults and hold between them a large piece of California—from Bakersfield to Redding, roughly, and including San Francisco, Sacramento, and Fresno—not to mention the whole of the High Sierra, Reno, and ten million acres of Nevada. "You create a California Plate," he says. "And the only question is: Is it this size, or the larger one? How much goes out to sea?" British Columbia is to his left and Mexico is to his right, beside his coffee cup on the oak Formica. The coast is against his belly. He moves his hands as if to pull all of central California out to sea. "Does this much go?" he says. "Or do the Mojave and Baja go with it?" A train of flatcars pounds through town carrying aircraft engines.

My mind has drifted outside the building. I am wondering what these people in this dry basin—a mile above sea level—would think if they knew what Deffeyes was doing, if they were confronted with the news that an ocean may open in their town. I will soon find out.

"What?"

"Are you stoned?"

"The way I see it, I won't be here, so the hell with it."

"It's a little doubtful. It could be, but it's a little doubtful."

"If it happens real quick, I guess a couple of people will die, but if it's like most other things they'll find out about it hundreds of years before and move people out of here. The whole world will probably go to hell before that happens anyway."

"You mean salt water, crests, troughs, big

splash, and all that? Don't sweat it. You're safe here —as long as Pluto's out there."

"We got a boat."

"That's the best news I've heard in a couple of years. When I go bye-bye to the place below, why, that water will be there to cool me. I hope it's Saturday night. I won't have to take an extra bath."

"It may be a good thing, there's so many politicians; but they may get an extra boat. I used to be a miner. Oh, I've been all over. But now they've got machines and all the miners have died."

"The entire history of Nevada is one of plant life, animal life, and human life adapting to very difficult conditions. People here are the most individualistic you can find. As district attorney, I see examples of it every day. They want to live free from government interference. They don't fit into a structured way of life. This area was settled by people who shun progress. Their way of life would be totally unattractive to most, but they chose it. They have chosen conditions that would be considered intolerable elsewhere. So they would adapt, easily, to the strangest of situations."

"I've been here thirty-three years, almost half of that as mayor. I can't quite imagine the sea coming in—although most of us know that this was all under water at one time. I know there's quite a fault that runs to the east of us here. It may not be active. But it leaves a mark on your mind."

"Everybody's entitled to an opinion. Everybody's entitled to ask a question. If I didn't think your question was valid, I wouldn't have to answer you. I'd hope the fishing was good. I wouldn't mind having some beach-front property. If it was absolutely certified that it was going to happen, we should take steps to keep people out of the area. But as chief of police I'm not going to be alarmed."

"It'll be a change to have water here instead of desert. By God, we could use it. I say that as fire chief. We get seventy fire calls a year, which ain't much, but then we have to go a hundred miles to put out those damned ranch fires. We can't save much, but we can at least put out the heat. I got a ten-thousand-gallon tank there, which is really something for a place with no water. I guess I won't still be here to see the ocean come, and I'm glad of it, because I can't swim."

Meanwhile, Deffeyes, in Sturgeon's Log Cabin, applies the last refining strokes to his sketchings on the map. "The Salton Sea and Death Valley are below sea level now, and the ocean would be there if it were not for pieces of this and that between," he says. "We are extending the continental crust here. It is exactly analogous to the East African Rift, the Red Sea, the Atlantic. California will be an island. It is just a matter of time."

THE PALEOZOIC ERA
340 MILLION YEARS

MILLIONS OF YEARS BEFORE THE PRESENT	System / Period		Stage — EUROPE	Age — NORTH AMERICA
230				
	PERMIAN		TATARIAN	OCHOAN
			KAZANIAN	GUADALUPIAN
			KUNGURIAN	
			ARTINSKIAN	LEONARDIAN
			SAKMARIAN	WOLFCAMPIAN
280				
	PENNSYLVANIAN	CARBONIFEROUS	STEPHANIAN	VIRGILIAN
				MISSOURIAN
			WESTPHALIAN	DESMOINESIAN
				ATOKAN
310				MORROWAN
	MISSISSIPPIAN		NAMURIAN	CHESTERIAN
			VISEAN	MERAMECIAN
				OSAGEAN
345			TOURNAISIAN	KINDERHOOKIAN
	DEVONIAN		FAMENNIAN	CHAUTAUQUAN
			FRASNIAN	SENECAN
			GIVETIAN	ERIAN
			COUVINIAN	
			EMSIAN	ONANDAGAN
			SIEGENIAN	ORISKANYAN
395			GEDINNIAN	HELDERBERGIAN
	SILURIAN		LUDLOVIAN	CAYUGAN
			WENLOCKIAN	NIAGARAN
			LLANDOVERIAN	
435			ASHGILLIAN	MEDINAN
				CINCINNATIAN
	ORDOVICIAN		CARADOCIAN	TRENTONIAN
			LLANDEILIAN	BLACKRIVERAN
			LLANVIRNIAN	CHAZYAN
			ARENIGIAN	CANADIAN
			TREMADOCIAN	
500			DOLGELLIAN	CROIXIAN
			FESTINIOGIAN	
			MAENTWROGIAN	
	CAMBRIAN		MENEVIAN	ALBERTAN
			SOLVAN	
			CAERFAIAN	WAUCOBAN
570				

PRECAMBRIAN TIME

ABOUT 4000 MILLION YEARS

MILLIONS OF YEARS BEFORE THE PRESENT	System Period		Regional Time Scales	
			EUROPE	NORTH AMERICA
570				
	HADRYNIAN		TORRIDONIAN	KEEWEENAWAN
1000	HELIKIAN	NEO-HELIKIAN		
		PALEO-HELIKIAN	LEWISIAN	HURONIAN
2000	APHEBIAN			
2600				LAURENTIAN
	THE ARCHEAN EON			KEEWATINIAN
4600				